Williamina Fleming, Astronomer

An Imagined Memoir
Susan Vizurraga

Gruff Brothers Books

Contents

Cambridge, 1905

So.
They say my name
(Williamina Paton Stevens Fleming—
Mina, if you please)
has been put forward
again
for this coveted prize in Astronomy.

But they also say that
I'm an unlikely choice,
being a woman
and therefore not a fellow,
not eligible to be named
a member of their august Academy.
(My discoveries of stars,
many of them,
and the order that I have put to their multitude,
apparently notwithstanding.)
I understand, of course.

But dinnae you think for one moment
that these respected, learned men
could have spent their long days
peering through their magnifiers
at their glass plates and the stars photographed there,

making their own computations,
preparing their own discoveries for press,
delivering their eloquent lectures—
all while managing a household,
raising a bairn into a fine young man
(an engineer, no less!)
and bearing all the burdens and worries
that fall to a woman,
and on a woman's wages, to boot.

Dinnae you believe
these grand men
with their prestigious degrees,
their mansions and their clubs and their cigars
—and their wives—
could ever have done all that!

Och.
No.

My apologies.

I should not have said so.
I am grateful,
fortunate,
I am,
to have been in their company—

especially considering
where I began.

Boston, 1879

I return,
later than I hoped, to our rented rooms,
two flights of narrow stairs
up from the muddy street,
relieved to find that James is not at home.

I have been to the butcher
for a small joint of beef,
to a pushcart near the wharf
for the cheapest potatoes, carrot, onion.

Supper will be a celebration.
And an apology.

I spoke too harshly, last night,
counting our next month's rent
from our savings,
laying out the few remaining notes and coins,
making plain what he surely already knew:
he must find work.
We cannae wait
for the position that he had expected at the bank.

And then . . .
Och.

And then,
last night,
I told him
what I had suspected for weeks—
that there will soon be a bairn to support, as well.

Aye,
the news should have waited for a better time,
a time for celebration
rather than trepidation.
But I was carried away
by worry.

I stoke the fire in the stove,
salt the beef, peel and cut the vegetables.
The roast is James's favorite,
a peace offering.
It will show him I am confident
that better times are ahead.
Though he is older than my twenty-one years,
he is yet young enough
to make a fresh start.
I winnae tell him
how careful I was with the shopping,
how I searched today for cheaper rooms,
how I asked at the tailor's whether they might need a seamstress.
(They did not.)

When supper is ready
and James has still not returned,
I sit by the stove,
arrange in front of it the damp, muddy hem of my dress to dry,

then brush away the dirt.

James is still absent
when I finally eat a forkful of beef and potato,
cover the pot,
bank the fire,
undress for bed alone.

James had complained,
when we first rented these furnished rooms,
of the narrowness of the bed.

Tonight,
as I tuck the quilt around my shoulders,
the bed feels much too wide.

Morning, 1879

I am still alone
when I wake to the cold morning light.

I lift the quilt from the bed and wrap it round me,
warm my feet and hands at the still-hot stove,
pull my wool stockings from the clothesline strung over the draining board.

Where can he be?
I thought sure he'd return in the night,
whispering promises to take whatever work he could find,
laying a hand on my expectant belly,
assuring me that all was well.

Has he gone wandering too far,
turned to a friend,
found a bed in a tavern,
so angry or disappointed in his circumstances
that he cannae face his wife?
Or—

Och.
Or is he somewhere in a Boston back street,
robbed or injured,
lying in the frozen mud of horse and cart tracks?

I throw off the quilt and dress quickly in the cold of the bedroom,

rolling on my stockings,
tying drawers and fastening corset over chemise,
tying on underskirt and bustle,
pulling yesterday's dress over it all,
last squeezing my feet into the shoes
that have been drying in front of the stove.

I take a crust of bread and dip it in last night's gravy,
something in my stomach so I'll not feel faint.

I will search for him,
though I scarcely know where to start.

Discovery, 1879

I hurriedly pin up my hair without brushing it,
fix my hat in the small mirror over the washbowl.

I may need money for whatever has befallen James, so
I pour out again upon the table
the too-light envelope of our finances.

It is gone.

Not all,
but I quickly recount the remains of our savings,
add the few remaining coins, still in my pocket,
from yesterday's groceries.

Half.

Half of what was there
when we argued the night before last.

I sit,
suddenly unable to stand.
He cannae have done it.
James cannae have left me.

I find myself in the bedroom,
afraid to look for his valise under the bed.

But I do look,
and I do not find it.

And the cupboard—
his extra shirts, his good stockings and trousers and underclothes
—all gone.

Now I notice the absence
of his razor by the washbowl,
his overcoat behind the door.

The absence of my husband.

One Day of Despair, 1879

Stunned,
I do not remember unpinning my hat,
laying my clothes upon the bed,
crawling under the quilt.

I turn toward the wall, regretting our last words to each other.
I stare at the ceiling,
imagining the perils of pregnancy, of giving birth alone,
abandoned by my husband,
an ocean away from my mother.
There is no one here to help me.

I curl under the quilt, fearing the months to come,
of begging for work with a growing belly
and later a bairn in my arms.

I do not mark the passing of time,
but I sleep, and wake, and turn to the wall
to start again the cycle of despair.

A Day of Determination, 1879

I wake much later in the dark, in the lonely bed.
I light the lamp and the cold stove
and, noticing the pair of James's thick wool socks
left forgotten on the clothesline,
I pull them on to warm my feet.

I am famished.
Wrapped in the quilt, I sit at the table
eating bread and butter by lamplight.

When the stove is hot
and the roast and gravy have warmed over again
I eat a bit of that, as well,
then place the rest in a covered bowl on the north windowsill
where the shade and the weather will preserve it.
I cannae afford to waste,
so it must be my supper for the next several days.

I cannae afford, either, another day of despair.
I cannae afford to wish
for the return of someone who would leave his pregnant wife
without a word.

I have less than a fortnight paid ahead on these rooms.
I must eat, wash, dress.
I must find a way to carry on.

Beginning Again, 1879

An ocean away from Scotland and
alone
except for the stirring in my belly,
a bairn-to-be I'd
—we'd—
hoped for once
(I had thought that, at least,
but now, abandoned, I see my mistake.)
I'll take what decent work I can find,
and quick,
for a roof and a meal.

Climbing Observatory Hill, 1879

I have asked at every store front, every office, in Boston.
I have offered myself
as bookkeeper, seamstress, milliner, tutor—
all work I have done, of a fashion,
and would be content to do again,
if it would provide a means of living
and a means of eventual improvement.

Today
on the advice of a few kind acquaintances
I spend precious pennies on a carriage to Cambridge
to do the same there.

We are crowded into the carriage:
a man opposite me whose stares I ignore,
another who coughs incessantly into a handkerchief,
beside me an impatient woman
who clicks her tongue at them both.

She tells me, pointing,
as the driver helps us down at the village common,
that there are new homes around Observatory Hill,
some perhaps in need of domestic service.

Domestic service.
It is not what my mother wished for

when she struggled to keep me in school.

I sit on a bench to get my bearings,
breathing in the cold air after the stuffiness of the carriage,
looking over the green of the common
and the war cannon displayed there.
It is not only the cold air
that causes me to press my handkerchief to my teary eyes.

I recall the story my grannie would tell us,
when we were weepy or afraid:
how we would have lived in a castle in Stirling
if only my great-great-grandmother had been a son to the laird
instead of a daughter.
How that laird's daughter had followed her husband
to fight Napoleon in the Peninsular Wars,
had seen her husband killed on the battlefield,
had given birth to their son on that very day,
on that very battlefield.
That son, she would say, was my grandfather,
Grannie's dear husband.
"You are made of that same stern stuff," she'd say,
checking the shadows for our imagined ghosts,
drying our eyes and applying plasters to our skinned knees.

Domestic service.
If this is what I must do for myself and my coming child,
I shall do it.
I stand, smooth the front of my dress as best I can,
and begin my walk up the hill.

A Roof, 1879

I'm to be a maid
to Director and Mrs. Pickering in Cambridge.

Aye, I have come to this—
a servant.
I had never pictured it.
I had planned to return to my studies,
as soon as James had established himself.
But now,
after two weeks of searching for a suitable position,
with all my worldly possessions in a trunk and a valise,
and the meager remains of our savings—half of it—
in a tattered envelope in my bodice,
I will be a maid.

But the Pickerings are gracious,
the Director accepting my references, detecting, I suspect,
that while I am unaccustomed to domestic service,
I'll not be ashamed of it.
And Mrs. Pickering assessing my appearance, detecting, perhaps,
my desperation, perhaps even my condition,
though she seems too discreet to remark upon it.

And it is a grand home,
polished floors reflecting gaslit sconces, rich draperies and carpets.

But best of all,
it is a part of the College there,
connected, quite literally,
to the Harvard Observatory.

It is more than a home,
and the roof over my head tonight is more than shelter.
It is a window to the stars.

A Meal, 1879

I am second maid,
there already being an Irishwoman who cooks and cleans.
She sets me to polishing silver,
dusting the books in the library,
gathering the bed linens to be taken to the laundress.
I fear there is not enough work to keep me
and that I have been taken in for charity
in what Mrs. Pickering calls "a delicate condition."

She is kind, to be sure.
But I'll yet have a need for employment
when the bairn is born,
provided James does not repent and return to be the husband
he vowed to be when we stood together
at the Church in Dundee just two years past.
How charitable might Mrs. Pickering still be then?

I help the first maid
(Hannah, a capable woman of twice my age,
only slightly offended to find me in her kitchen)
with dinner
and while I whisk and smooth the gravy
I offer, with her permission, to make fudge from the provisions in the
    pantry.

"Suit yourself," she says,

"I'm off after serving dinner. Just don't leave your mess for me."

It has just the time to cool and be cut
when Mrs. Pickering calls for coffee,
and I offer the candy on a paper-lined tray
to Director Pickering and their guests, a Professor of Physics and his wife.

"Mrs. Fleming!" Mrs. Pickering exclaims.
"This is just the thing for my teas with the students.
Could you manage, I wonder, a batch every Friday?
For a dozen? Best make extra, in fact," she laughs to her guests.
"The young men are away from home and always hungry, I find!"

"Certainly, Mrs. Pickering," I nod.

"Splendid!" she says, "You'll tell me the recipe tomorrow, and I'll figure
enough ingredients and change the order for the grocer."

"Oh, no need," I say, "I'll just add thrice each to the order
that Hannah has ready."

"Well." Mrs. Pickering casts a discerning look at me
and I fear for a moment that I have caused offense.
"Excellent," she says.

Later,
sitting in the kitchen with some bread and gravy,
I am more than pleased to have proven myself useful.
And before I wash the last of the dishes,
I enjoy a cup of tea and, finally, the fudge
that, for a number of trying days,
I have so craved.

Observatory, 1879

I had thought of a telescope
as the toy my brothers held to their eyes
while playing pirate on the docks at Dundee,
searching the horizon for ships and sails and shapes in the storm clouds.

But here in the Harvard College Observatory
there is a great dome
and inside it a telescope large enough to dwarf a man,
a machine of brass and gears and glass,
polished wood and iron rails,
a chair like a sleigh, and cogs and pulleys and weights,
a sliver of the concave ceiling
sliding open to reveal a slice of the clear night sky.
What a wonder it is!

Novae, 1879

"Join us, Mrs. Fleming," Director Pickering suggests.
"See what pays our bills here at the Observatory."

"Only some of the bills, my dear," Mrs. Pickering smiles.

Their guest tonight, a Latin scholar, finishes his brandy.

I gladly follow through as they leave the residence through Director
    Pickering's office,
past clocks and tools and telegraph wires and a machine,

"A Meridian Circle," he explains, "to observe stars as they cross the
    meridian. And," he adds, "a most accurate timepiece. We telegraph the
    time of day to the railroads. They set their clocks by it."

"Ah," says the Latin scholar, smiling, "literally making the trains run on
    time."

"Precisely. Here we are," Director Pickering says, holding open the door to
    the Great Dome. "The Great Refractor."

A young man, one of the Director's assistants, goes on with his work,
laboriously turning wheels that adjust gears and move the telescope into
    position.

"And now to discover stella novae!" the scholar exclaims, climbing into the
   intricate observing chair.

"Is it possible, new stars tonight?" I ask.

"Ah. If there is a star we don't recognize," Director Pickering explains, "it
   is not until we examine our records and continue to observe that we can
   actually claim a discovery."

"But where is the romance in that?" the scholar interrupts.

Mrs. Pickering laughs. "You can always hope for a comet, Victor," she says.

"Do women assist in your work, Director Pickering?" I ask.

"With our calculation, at times," he answers. "We have a few women
   computers, unusually adept at mathematics. But not in the Observatory
   proper. The machinery is heavy and requires some strength to manage,
   as you can see. And of course it is quite cold with the dome open in the
   winter months."

I nod,
but later as the Director leads us back toward the residence
I find myself thinking
that perhaps he should acquaint himself
with the weight of the heavy iron cookpot that Hannah swings off of the
   stove
and the temperature of the wind gusting across the Firth of Tay
in a cold Scots winter.

Daguerreotype, 1879

I take the feather duster to the parlor mantel
of the Director's residence,
where the coal dust tends to settle.
I wind the mantel clock
and lift to brush, one by one,
the Daguerreotypes displayed there:
one the Director himself in academic regalia,
another a younger Mrs. Pickering, I believe,
another a wedding portrait of the Pickerings,
others whom I do not know.

I see myself in the mantel mirror,
and imagine my reflection as a young girl,
at my own father's studio above his shop in Dundee,
in the silver-coated copper plates
that he polished to perfect mirrors.
When a grand man or a fine lady
would come to sit for a portrait,
he would arrange them just so,
and allow me to observe
as he would quickly cover a mirrored plate
with a yellowing coat of chemicals,
set it smartly into its hinged wooden box
and slide it with a click into his camera.
He would disappear, then,
under the black velvet drape,

his muffled voice warning the gentleman or lady
(or myself, as well, if I were dancing about)
to be still
while he focused the lens
and then while the light impressed itself
upon the prepared plate.
Then he would have me escort the sitters
back down the stairs to the shop,
while he opened the windows to the breeze and
finished and fixed the image with mercury vapor,
trays filled with salted water,
rinses of gold salts,
and the heat of a flame.
The images would be preserved, then
as these on the Pickerings' mantel,
fitted behind glass
in tight frames that folded,
I always thought,
like pretty little books.

How I wish now
that I had an image of my own father,
like those he made for his patrons!

There is an opening door,
a swish of silks,
and Mrs. Pickering enters the room,
a vase of flowers in her arms.
"Oh, Mrs. Fleming," she says,
"do be careful with those.
A cracked glass could damage the image
by exposing it to air."

"Oh, yes, Mrs. Pickering," I say,
gently arranging the collection again.
"My father had a shop in Dundee.

I used to watch them being made."

"Ah," she says,
as I move on to the windowsill,
catching a cobweb with the feather duster.
"There is more to discover of you
than one might suspect, Mrs. Fleming,"
she laughs,
and she sets the vase of flowers atop the mantel,
arranges the stems,
brushes her hands together.
"But now I must dress for dinner," she says.
"There will be eight tonight.
Will you please see to the dining table
when you finish here?"

"Of course," I say,
and I leave my memories
for another day.

Alterations, 1879

I have tied my drawers and underskirt less tightly,
loosened the laces of my corset.
Now, lighting the lamp in the little room off the kitchen where I sleep,
I sharpen my sewing needle,
cut a length of thread,
and begin resewing my spare work dress.
I have carefully opened the seams,
heated the iron,
and pressed a narrower fold into the edges,
letting out all that I could
—nearly an inch, I think—
to accommodate the changing shape of my body.

I examine the dress in my lap.
Perhaps I can move the buttons a bit?

The pinny I wear while I do my work
can hide me for only so long,
and though my needlework will do for a while,
I will need to buy fabric
and sew a new dress, or two.
And then
there will be need for baby clothes,
and flannels for diapers, and
och,
there will be need for so much.

I rub my eyes,
tired from the fine work
in the dim light.

Did James bother himself, I wonder,
to think of this
when he counted out this half of our "fortune"
and left without a word?

And what could I ever have meant to him, truly,
if he could have done that to me?
To us?

Society, 1879

The Pickerings' guests tonight
are a few of the members of Boston Society,
Hannah says,
from the finest homes,
from families who have led the city for more than a hundred years.
So I am surprised,
when I carry in the coffee service,
to encounter talk
not of politics or economics,
but on the recitation of poetry.

"And here is Mrs. Fleming," the Director says,
as I set the tray upon the table.
He addresses his guests.
"We have just had Longfellow.
Perhaps now Robert Burns,
in honor of her countryman?"

"I believe I can oblige," says one of the guests,
rising and turning toward me.
"O, my luv is like a red, red rose," he begins,
in what is a rather terrible imitation
of a Scots accent.

"Do you know it, Mrs. Fleming?" Mrs. Pickering asks.

The gentleman continues, and motions for me to join in.
Perhaps I should modestly shake my head
and remove myself to the kitchen,
but the company seem friendly and jovial,
and I cannae resist.

The gentleman has quieted and left me to finish it alone,
". . . and I would come again, my luv,
though it were ten thousand mile."
There is a round of applause and laughter.

"Well done, Mrs. Fleming," the Director says.
"It seems we are still discovering your talents."

"Och," I say,
smiling as I remove the dessert plates,
"I am a Scot, Sir, after all."

The Garden, 1879

"Will you help me in the garden, Mrs. Fleming?" Mrs. Pickering asks,
tying on a pinny
and arranging her hair beneath a brimmed hat.

I glance toward Hannah who "tsk"s me away.

"Take a pail for the spent flowers," Mrs. Pickering says,
"and a hat, if you wish.
although I'd think, with your coloring,
you're not so much at the mercy of the sun."

Although it is true
that my hair and eyes are dark,
I do try to keep my face from the sun.
But I have only the one hat at present,
and it is more fashion than shelter.
I dinnae wish to see it soiled.

In the garden on the Observatory grounds,
Mrs. Pickering sets me to snipping the spent heads of lilacs
while she unrolls a mat of waxed canvas and kneels to pull at weeds.

"The gardens are lovely," I say.

Mrs. Pickering smiles brightly.
"I have been told," she says,
"that they rival those of the Arboretum.

But they do require much of my attention."

"I would have thought," I say,
"that the College would send someone to tend to the gardens."

"Oh, they would," Mrs. Pickering says.
"And that would no doubt find its way into the Observatory's finances.
My work keeps the costs down, you see,
and that helps the Director.
Why," she leans toward me conspiratorially,
"Edward even sells the grass clippings,
all to finance the Observatory's work.
And I can cultivate what I like.
Did you have a garden, Mrs. Fleming?" she asks.

"I did not," I say.
"I was raised in the city, where my father had his shop.
He made gilded picture frames,
as well as studying the work of Mr. Daguerre."

"Ah, yes. And does your father continue this work?" she asks.

"Och, no," I say.
"He died when I was seven.
My mother tried to continue, with the help of my older brothers,
but it couldn't be done, with the younger children to care for.
She had to sell it all away."

"What a tragedy!" Mrs. Pickering says.

"For a time, it was," I say.
"But she was determined that we all be schooled,
although she could have sent the boys to work on the docks,
and me to work at the marmalade factory.
Instead, I continued in the school,
until I was judged capable,

at fourteen,
to become a teacher myself."

"At fourteen!" Mrs. Pickering exclaims.
"Until you married, I suppose?"

I nod.
"I had thought to study bookkeeping,
as Mathematics were always my favorite. But."
I set down the pail and take a moment to rub my back,
and my condition makes the rest obvious.

"But another fate awaited you," Mrs. Pickering smiles consolingly.

"Have you had children, Mrs. Pickering?" I ask.
This is the way we ask each other,
one woman to another,
knowing how many babes do not survive infancy, or childhood,
though they never pass from love or memory.

Mrs. Pickering sits back on her heels.
"No, that is one way in which we have not been so blessed."

"Well, perhaps . . ." I begin,

but she interrupts.
"I have been told not to expect it," she says firmly,
and turns back to her work.

I cast about for what to say next.
"And so," I say brightly,
"you can devote yourself to creating your own Eden here."

"Yes," she smiles.
But a moment later she jumps up, startled.
We both watch as a thin green snake winds away through the grass.

Mrs. Pickering brushes her gloved hands.

"Eden, indeed," she says.

Useful, 1879

"Useless!" the Director mutters,
entering through the kitchen from his office in the Observatory,
thrusting a sheaf of papers upon the kitchen work table.
"Such a lack of . . . of industry!
Of . . . of . . . responsibility!"

I turn, disturbed,
from the pantry
where I have retrieved the canister of sugar.

"Oh, my apologies, Mrs. Fleming," he says.
"A bit of frustration with my assistants, I'm afraid.
The young men are happy to spend their evenings at the telescope,
but must be coerced to write their observations
in any decipherable form, and to do the necessary reductions the next day.
And they show little interest in the photographic plates!
Quite infuriating, at times."
He rubs his temple and takes the papers back in hand.
"And I am late to supper," he says
in a voice considerably calmer and a few tones lower.
"I am sorry to alarm you," he says,
and retreats through to the dining room
where his wife awaits.

Later, while I am listening for the kettle to boil,
I overhear the conversation

that accompanies the dessert pudding at the dining table.

"Did you know, my dear," I hear Mrs. Pickering say,
"that our Mrs. Fleming is adept at mathematics,
reads a bit of Latin, and spent her childhood in the studio of a
    Daguerreotypist?
Do you not think that her talents might be put to higher use?"

Still later,
when I enter to pour more tea,
the Director indicates an empty chair at the table.
And when I cautiously take a seat,
he suggests a change to my employ.

Transit, 1879

If one could fly
through the ceiling of an upstairs room of the Director's residence,
up and over the Observatory dome,
and float down, somehow,
through the corresponding roof of the other half of the building
in which much of the true work of the Observatory is done,
one would be in the East Computing Room.

After my morning's work in the residence
(after Hannah is satisfied with the tasks she's assigned me)
I leave behind the broom and mop and brush,
take off the pinny marked by dust and dishwater and shoe polish,
and enter the East Computing Room,
where there are journals with marbled covers and lined pages bound with
    string,
loose leaves of paper in neat stacks upon the shelves,
rulers so that our lines may be straight,
an assortment of pencils, pens, and inkwells,
bottles of ink in black, blue, and red,
leather-bound volumes filled with maps and charts and catalogues of the
    stars,
the astronomical knowledge of the world.

Och, I dinnae know why I describe it as flying.
My shoes are firmly on the ground,
my skirts and petticoat never higher than my ankles,

my sleeves securely buttoned at my wrists.

But as I transit each day
from the residence side to the scientific side of the Observatory,
from the work of a maid to the work of a computer,
it seems my step is quite a bit
more light.

In the East Computing Room, 1879

I join four other women:
Mrs. Rogers, wife of the astronomer,
Misses Winlock and Bond, each being a daughter of a previous
    Observatory Director, and Miss Bond the sister of another, as well,
and Miss Saunders, placed there by Harvard's President Eliot himself.

They are women who,
though limited in education and profession by their sex,
have been long immersed in the sciences,
learning by assisting fathers, brothers, husband.

I have no such experience,
no credentials to match,
but they are gracious and welcoming, clearing a space at the end of a table
    for me.

And as I watch them
deciphering the astronomers' notes scribbled in the dark of the night
    before,
applying the formulae that will accurately fix each star to a slice of the stellar
    map,
I feel more suited to their company,
more suited to this room,
more suited to this work,
than to any I have ever been before.

New Work, 1879

At first I am merely a copyist,
reproducing with pen and loose paper
the announcements and astronomers' remarks and tables of figures that are
    too dear to be lost when sent to the printers,
copies of letters the Director is sending to scientists and professors and
    industrialists, so that he will have record of them in his files.

I also become an accountant,
calculating the hours and pay of the professors' assistants and the
    computers,
figuring the costs of office supplies,
the coal for the fires and gas for the lights,
the parts required to keep the equipment in repair,
preparing cheques for the Director's signature.

And then,
because there is so much of the work to be done,
I become a computer.

Miss Bond shows me her work.
"Do you believe you can manage it, Mrs. Fleming?" she asks.

"Aye," I say. I believe I can.

Computation, 1879

Now,
as well as copying and accounting, I join in the computing.

I am familiar with the numbers, the Greek letters, the super- and
 sub-scripted numerals.
I can decipher what to do with the values in parentheses, the roots and
 squares, the functions and variables.

I do not yet know how the formulae are derived, but I will learn.

To be exacting in the calculation of the positions of the heavenly bodies
as they meet the crosshairs of the telescope over the millions of miles,
I had never before considered the necessity of mathematical formulae to
 account for
the bend of light waves through the atmosphere,
the Earth's place in its journey around the sun,
the tilt of its axis,
its very curvature.

Respect, 1879

I do not have the pedigree of the other women computers.
I have no astronomer, physicist, mathematician, philosopher in my family
    tree.
I've had no scholars of science to tutor me.

But I had a clever and curious father,
a wise and steadfast mother.

And though my husband has abandoned me
I have the determination to make my way.

And now, though it is only weeks that I have done this work,
with these women,
now,
I have their respect.

An Imprecise Calculation, 1879

The formula, such that it is, for my condition,
is not in thousands of miles nor in hundredths of degrees,
but in months. Weeks. Days.

It is quite simple.
But the body and the memory
cannot be so relied upon
as the astronomical tables.

I dinnae know precisely when the bairn will come
and I dinnae know precisely what I will do when it does.

A Word, 1879

"A word, Mrs. Fleming," Director Pickering requests,
"when you finish here, if you have a moment. In the residence, perhaps."

It is half-five
and I am so struck with fear that this is the end of my employment,
that I will be done in by this pregnancy
just as I have become adept at the work of a computer,
that the numbers float away in my mind.
I find it impossible to calculate,
and spend the last thirty minutes of my time
arranging the many notes and journals, refilling inkwells, sharpening
    pencils.

At six, I collect my hat
and take the journal, a ribbon holding the place of my last computations
so that I may complete my duties in my room,
and make my way through to the residence.

The Director meets me in the hall and invites me to the parlor
where earlier today I had dusted the tables and desk, brushed a spot of dried
    mud from the carpet.
Mrs. Pickering looks up expectantly as the Director indicates a chair for
    me.

"Well, Mrs. Fleming," he begins,
"I am most pleased with your work

both here in the residence and in the computing room. But . . ."

Here I grip the arms of the chair as the thoughts gallop through my head:
What will I do?
What will I do with a bairn and no income?
Will I impose on my poor brother and his young wife in Boston?
Return alone to Dundee
and hope to teach again in the school next to the marmalade factory,
the children's attention drawn to smells of orange and burnt sugar instead
    of their lessons?

"But," the Director continues,
"you are far too valuable as a computer to continue, even part of your time,
    as a housemaid.
I would like," he says, "for you to work all of your time at the Observatory.
There you will be an employee of the College, and not of this household."

He looks at Mrs. Pickering, who smiles and continues herself.
"I've told Edward that you will need time, however, for . . . your
    confinement.
We thought, Mrs. Fleming, that you might be considering returning to
    Scotland,
to your mother, for help with the childbirth . . ."

"And that," Director Pickering continues, "when you're quite ready,
you might return to continue your work on the other side, in the
    Observatory.
Would that be satisfactory?"

I am dumbfounded
and try to prepare my thoughts to speak.

"If so, Mrs. Fleming," Mrs. Pickering takes up the conversation,
"Edward and I thought we might help with the cost of your passage to
    Scotland,

as we know your pay here has not possibly been sufficient enough to allow
    for it."

"That is," the Director takes another turn, "if you plan on returning to us.
To the Observatory.
The women computers, you see, Mrs. Fleming,
have proven so successful, and at relatively little expense,
that I hope to hire more of them,
if subscriptions and gifts to the Observatory continue to fund us as they
    have.
I would hope you would be one to continue the work here."

They both look at me expectantly,
and I am so nearly overcome
that I speak before I can compose a more dignified or formal response.
"I dinnae know what to say," I begin,
and I notice their smiles at the Scots in my voice,
"except to thank you, both of you.
Yes, I should very much like to continue my work in the Computing
    Room."
I stand, my hat and journal in my hand.
"Very much, indeed."

Director and Mrs. Pickering hasten to stand.

"Och," I laugh, shaking the hand of each,
"and I thought you were letting me go!"

The Observatory Pinafore, 1879

I have not yet been to the opera,
but Miss Saunders so enjoyed "The H. M. S. Pinafore" that she has bought
    the sheets of music,
and performed some of Gilbert and Sullivan's delightful work at our last
    social gathering.

Now Mr. Winslow Upton, one of the assistant astronomers, has written
    his own
"Harvard Observatory Pinafore,"
a clever parody to be performed, alas, in the coming year
when I will be away for my time in Dundee.

There is an astronomer in it, who sings to the tune of "Buttercup,"
"I'm called an astronomer, skillful astronomer,
Though I could never tell why,
But yet an astronomer, happy astronomer,
Modest astronomer, I.
I read the thermometers, break the photometers,
Mend them with paper and wax;
I often lament that so seldom is spent
A fair evening on star parallax.
I write many letters, give aid to my betters,
and often sit up late o'nights
To catch a few glimpses of the many eclipses
of Jupiter's bright satellites."

There is our Director Pickering, the "gallant Captain of the Observatree"
who moves among "men of note" but "hardly ever wears a swallowtail coat,"
and there is a "Chorus of Computers," of which I would be a part, who
    sing,
"They work from morn 'til night,
For computing is their duty,
They're faithful and polite,
Their record book's a beauty."
So true!

Mr. Upton asks me to write a copy of his draft,
and points out his amusing reference to "my" character, a "Scotch maid,"
who would teach the "Chorus of Computers" to dance the Highland
    Polka,
if only she were not gone back to her native land.

I may not be at the "Observatree" when Mr. Upton's musical is performed,
but it is pleasant to know that I will not be forgotten.

Preparations, 1879

It will be late September, perhaps early October,
as best I can calculate,
when I will, God willing, be delivered.

I will lie in
at my mother's house back in Dundee.

Under steam and sail I will be at least two weeks at sea.
Poor winds and weather could add days—weeks—more.

Now that I am decided, I must make haste,
repack the trunk that James and I brought with us barely half a year ago.
Now without a husband's clothing and sundries
there is room in it for needles, thread, and sharpened scissors,
the pieces for a baby's dress already cut,
a doll for which I'll fashion clothes,
the fabric I bought in Boston so that I'll have a new dress when I return
to a real position at the Observatory.
I'll add dry crackers, some salt beef to keep something in my stomach to
    stave off sickness,
perhaps small squares of fudge wrapped in paraffin paper,
a small indulgence that I might look forward to each day of the journey,
and of course, for reading,
the latest publishings of the Observatory astronomers
so that while I give birth and care for a bairn
I'll not fall too far behind.

Journey Home, 1879

The journey will be coach and ship and train and carriage.

Mrs. Pickering is kind enough to accompany me on the coach from Boston
    to New York
where I have secured passage on a ship to Glasgow.
She will be calling on an heiress
with an interest in astronomy and a fortune to spend on it,
perhaps securing additional funding for the Observatory.
She spends much of the two-day trip rereading the writings
of the heiress's late husband, a respected amateur with a coveted telescope.

"They have no idea, do they, Mrs. Fleming?" she says,
"Men, that is, and what work women constantly perform
behind the scenes of their success."

As her husband is my employer, I demur.
But I place my hand on the front of my dress and agree—
of many things
men have no idea.

At Sea, 1879

I must either
lie on the berth in the dark close cabin I share between decks,
the odor of those around me and whatever is in the hold turning my
    stomach,
or sit above on the deck itself,
where I am exposed to the elements but at least have light and air.

Between them is a ladder
difficult enough in skirts
but worse in a condition in which I can scarcely see the rungs on which to
    step
and in which I seem, like the planets,
to wobble on my axis.

With either choice,
I cannae escape the rolling of the ship in the waves.
I cannae escape the rolling of the growing body within my own.

Scotland, 1879

Finally
the craigs rise from the sea
the cliffs close in
and I disembark, a bit shakily, in Glasgow.

Then to the train that steams east, then north, then east again,
windows open to air that smells of green,
a relief after the sea.

And a pleasant surprise, at last, in Dundee—
my brother Charles, now fifteen,
waiting at the station with a cart and horse
to carry me home.

Old Wives, 1879

"'T'will be a boy," my grannie says.
"Cannae ye see? The way she carries it low?"

"Aye," my mother says,
"but she's craving sweets, is she not? I only did that with my girls."

"Och," says Miss Margret Lindsay, the boarder
who has the spare room and whose rent helps to keep food on the table,
"hang a threaded needle over her belly, and see which way it sways.
Then ye'll ken for sure."

They speak as if I am not sitting before them,
more uncomfortable every day as my time draws close.

"And what will ye name this bairn, Mina,
be it boy or girl?" Grannie asks.

"I had thought to name it in honor of the Pickerings,
my employer and his wife. They have been so kind," I say.

"Do they not have children of their own?" Mother asks.

"None," I say.
"'T'would be Edward Charles or Lizzie—Elizabeth, I suppose."

"Best be sure," my mother says,

laying a hand on the front of my dress.
"I think ye dinnae have much time to decide."

Time, 1879

To bring on the labor
Mother suggests brisk walks outside despite the growing cold.

This evening I enlist Charles to accompany me along Alexander Street.

"What's it like, then, America?" he asks as we stroll.

"D'ye mean, besides the streets of gold?" I tease.

"Tsk," he says, "I mean the people, like.
The work. what could I do, there, d'ye think?"

"The people," I say, "are not so different.
As many Irish in Boston, I wager, as in Dublin. And English.
And American, however they came to be so . . .
But I have not seen so much.
And no cowboys," I tell him, remembering how he used to play.

He scuffs his toe on the street, impatient with me, so I go on,
"And work—so many chances to make your way.
But much is expected of you as well.
And what you should do, should you get there, Charlie," I say,
"is to study, to begin. No better place—"

But here I take his arm
because pain has gripped my body.
Then it passes.

"Home, then," he says,
and I nod, gasping.

It is not far
but it seems a long time as we make our way,
stopping every so often so I can lean, gasping,
on my brother's arm.

Sea of Pain, 1879

I change into my night dress.
Mother strips the bed and lays down a tarpaulin and a patched blanket atop
    it.
She rolls and folds blankets and props me up with them.

Grannie brings a cup of water.
"Only wee sips," she says, then goes to heat the pot on the stove
and sends Charles to dip more water from the barrel in the alley.

Mother sits in the chair beside the bed,
resting as she waits.

And I sit, lie, pace around the room,
the pain like a sea surrounding me,
lapping at my body,
and then, in waves, crashing over me,
pulling me under its surface.

Birth, 6 October 1879

It goes on and on, the pain,
but through the night and the morning and into the afternoon,
exhausted,
I am no closer to being delivered.

My mother and grandmother, no strangers to childbirth, nod to Charles
and I know they've decided
that he must fetch the doctor.

I attempt to shut from my mind the ones we've lost,
the bairns that came into the world
and left it without ever drawing breath,
the other women who did not survive the ordeal.

But then something turns to urgency
and Mother, so weary, brightens.
She rubs her hands, dips them in the basin
and dries them on her pinny.

Finally the pain is productive and,
just as Charles returns with the doctor,
I am delivered.

There's a cry—weak and then stronger—and then,
handing the bundle to me, the doctor says,
"Och, ye dinnae need me. Ye've a bonnie wee lad."

A Name, 1879

He must have his father's surname, of course,
by custom and law,
though I dinnae know what good that will do him.
Any honor in it will have to come from the lad himself,
or, perhaps, from me.

His given name is up to me,
and I mean to honor a better man.
My son will be christened Edward Charles Pickering Fleming.

Though it seems now a grand name
for such a wee bairn,
I will help him to grow into it.

The Atlantic, 1881

At the bow, holding tight to the railing
(I find I must always hold something in my hands,
else feel the lack of my son, behind in Dundee,
in the care of my mother and grannie.)
I watch the last arc of the sun disappear
as if into the sea.
Venus and the stars, no longer obscured by the sunlight,
make themselves known.

I think of the volume I found at the bookseller at Nethergate—sketches of
    the constellations—that I left with Charles,
making him promise to take my Edward out in the evenings
when the weather permits,
so that he and I might see the same stars.

For the Best, 1881

Aye, it is for the best,
to have Mother look after him
while I prepare a life for us here.

It is what many immigrants do—a sacrifice to be redeemed
when the leanest times are through,
when the nights in cold, bare rooms have passed,
when the food and streets and work have become familiar,
when the language is learned,
when the hardest edges are chipped away from the accent,
when whatever small funds finally have been set aside,
when families are reunited,

when, I hope,
it will all have been for the best.

A Women's Room, 1881

I am now the sixth woman in the East Computing Room,
a cozy nest I began to feather nearly two years ago.
"I trust you, Mrs. Fleming," the Director had said,
"to choose paint and paper, carpet, if you wish.
I want the women computers to be comfortable, as well as productive.
Desks, do you think?
Or tables, perhaps.
See to the orders, if you can, before you leave us."

Now, on my return, I note the result:
maps and charts and the likenesses of astronomers in their frames
look down from prettily-papered walls.
Shades let in enough sun to complement the gaslight.
Polished wooden shelves hold books and annals and photographic plates.
A carpet warms and muffles the heels, and the voices,
of the women at work.

Here I cannae wait to begin again.

The Computers, 1881

For 25¢ an hour, ten hours each day, six days each week,
we translate the hasty notes of the men at the telescopes,
reduce the equations that pinpoint the stars in the sky,
measure their brightness,
calculate their distances.
We consult the maps and the writing of the astronomers, the amateurs, the
    philosophers,
every one who looked to the heavens and recorded what he saw there.

After the days inside the Computing Room
with our notes and journals, our charts and pens and ink and papers,
Miss Saunders, Miss Winlock, Miss Bond, Mrs. Rogers, Miss Nettie Farrar,
    and I
step outside, walk down the hill
to our homes, our rooms, our waiting carriages,
while looking up at the sky.

Frugality, 1881

I live on very little
and save
so that I may send to Mother as much as I can
for the care of my son
and save
so that one day I may buy passage for them
from Scotland to America
and save
so that we might live
in a house more suitable than the rooms I rent now—
a home for a family, however small.

It will not be the home I had planned—
a husband, a wife, a houseful of children.
But a home it will be,
and in it, although there only be the two of us,
a family.

The Work, 1881

On most nights
Director Pickering or one of his assistant astronomers
centers the Great Refractor on a small portion of the sky,
fixes his gaze on a reference star, and notes every other body he sees,
assigns it a magnitude, and marks its position.

On most days
I copy the Director's notes,
apply the necessary formulae to each star, each position, each magnitude,
reduce and record the results in the tables of the Observatory journals,
copy and compute the astronomers' work to be included in the next
    Harvard Annals,
prepare Mr. Winlock's work for publication,
complete much of the Director's correspondence,
and keep the Observatory's accounts.

When I worked as a maid in the Residence,
I had hoped to make myself useful.
Working here in the Computing Room,
I take a wee bit of pride in knowing
that I am very useful indeed.

Missing Him, 1882

He was not yet two
when I set off back to America.
(Though the lad had already proved himself a keen one,
walking on his wee short legs
before he was a year!)

Now I can only imagine him
listening to his grannie's stories,
learning his numbers,
singing his Mother Goose rhymes.

I write letters to be read to him,
find picture books for him to look at,
sew clothes in increasing sizes.
I wrap them up with brown paper and twine
and send them across the sea.

They are poor substitutes,
but they are the only kind of love that I can send by post.

Missing Another, 1882

It brings on sadness, regret, confusion, fury.

So I do not think often of him,
the husband who abandoned me.

I still do not know his reasons,
and contemplation only makes me doubt myself,
a doubt I can little afford.

But I find sometimes I miss
the safety of a man beside me in public,
the authority that his existence conferred,
the prospects that his skills provided.

And sometimes, alone,
I allow myself to miss
his presence in the room with me,
his voice calling to me from another,
the strength and warmth and weight of him.

Photography, 1882

"Photography, ladies," Director Pickering exults,
"stellar photography—the future of the field of Astronomy!"

He holds a rectangular plate of glass,
a paint-like coating of white on one side,
which I know is an emulsion of chemicals
like that on the Daguerreotype plates in my father's shop in Dundee.
But this is not a finished portrait, but a glass negative,
a field of white splashed with pinpoint dots of black
where the light of stars has penetrated, an image in the opposite.
It is one frame of the night sky
recorded through the eye of the telescope.

"Soon you'll no longer be required
to decipher the nightly scribblings of the astronomer.
You will see what he saw—
and the record shall remain for future astronomers to discover!"

I have heard of the astronomical photographs
of the lunar eclipse, have seen the moon and its craters.
But the stars!
Imagine a photograph so fine that starlight is visible—
that it may be mapped, compared, observed over time.

And imagine that we, the computers and copyists—

who look only at the notes and drawings and conclusions of the men at the
    telescopes—
imagine that we, someday, might notice on one of these glass plates
the light of a star that no eye has seen before.
Imagine such a discovery!

An American Game, 25 November 1882

Harvard vs. Yale

Mrs. Pickering graciously hosts a luncheon for the Observatory "family"
and although she and most of the wives and computers stay behind after,
it is such a fine day
that I walk with some of the astronomers and assistants down the hill to
    Holmes's Field
joining two thousand or more (one-tenth, perhaps, are women)
where we sit on long benches and cheer on our boys.
They are calling it "football," but it looks little like the football I have seen
    in Scotland
and it seems to involve more throwing and wrestling than footwork.
I cannae deny that if my Dundee schoolboys had played this roughly at
    their recess
I would have sent them inside to exert themselves by scrubbing the slates
    of chalk.

The Harvard Eleven's noses and brows are bruised and bloodied
and I suspect their crimson jerseys disguise even more injury.

The rules are a mystery to me
(and, it seems, to many of those presently competing),
but I will learn.

I suppose to be the mother of a boy
is to be ready for games

and also for cuts and bruises,
for blood and mud and stains.
One day, my own boy
will no doubt wish to play.

The Score, 1882

Och, I forgot to note the result,
though it's likely best forgotten:
Yale, one goal and three touchdowns.
Harvard, nothing.
"Wait until next year!" they say.

But the game has given me an idea of a new plaything to sew and send to
    Dundee.
When I have an afternoon, I will look in the shops for an eighth of a yard
    of crimson wool
for a doll-size jersey and stockings, some white for breeches, an unbleached
    muslin for head, limbs, and body.
I'll work a yarn "H" onto the crimson jersey,
knit a miniature crimson wool cap
and send my Edward his own wee Harvard football player.

He is now three years old
and I ache to see him again.

Transit of Venus, 6 December 1882

Today
the second planet will cross, for a few moments, the face of the Sun.

It is not the romance evoked by the goddess of love and the flaming orb
that interests the astronomers.
It is the calculations that might be made
about the distances of the Sun from the Earth, Venus from the Sun, Earth
    from Venus—
about the size of the planet and of the Sun itself.

Each of the Observatory's telescopes, as well as those of amateur
    astronomers from Cambridge and beyond,
have been assembled on Observatory Hill.
There are makeshift screens upon which mirrors will reflect the image of
    the Sun,
small thick plates of smoked glass in the hands of those who will dare to
    peer more directly at the spectacle.

Venus will be invisible to our eyes
until it will gradually appear as a small dark circle—perhaps just a speck—
in front of our brilliant Sun.

We will call out the time intervals,
take note when the astronomers observe first "contact,"
when they observe that the whole planet is within the face of the Sun,
when it reaches the opposite edge,

when it has completely left its face.

There is a buzz of excitement, the checking of pocket watches.
Despite the cold, the Hill has the air of a picnic.

It is nearly two o'clock when we realize that the skies are not clearing.

No one sees the beginning of the transit.
There are a few readings when the clouds thin enough
to briefly make the dark mote of Venus visible.
But there will be no great discoveries today.

The best telescopes and lenses,
the most eminent astronomers,
the most ardent assistants.
Still, we are at the mercy of the weather.
A few clouds can make it all for naught.

A Visitor, February 1883

.

Mrs. Draper,
a charming and elegant woman recently widowed, visits the Observatory.
The Director shows her into the Computing Room, eager to show off the
    women at work there.
Some of us are at our calculations, our heads bent over the formulae,
and others at the light frames that have been built to study the glass
    photographic plates,
our eyes at magnifying glasses, studying the specks of stars,
calling out the readings to the recorders who write them down.

She shows us the tiny glass plates taken by her late husband,
starlight passed through a glass prism
that shows not the specks of stars, but the spectrum of light that they emit.
Not the colors of the rainbow, of course—no photograph can show
    color—
but a series of lines, distinct for each star,
that may hint at the unknown elements that are the sources of their light.

"As you can imagine, Mrs. Draper," the Director says,
"the Harvard Women Computers are more than capable
of helping with the necessary measurements and reductions.
I hope you will consider publishing the discoveries that you and Henry
    made.
Think, Anna, of the leap forward this could begin!"

Listening, I recognize the enthusiasm in the Director's persuasive words.

It is not just for the scientific leap forward,
but for the wherewithal that made it possible.
The Drapers' work, and their fortune, would be most welcome at the
    Harvard College Observatory.

Mrs. Draper seems reluctant, though,
and says she will attempt to continue on her own.

Och, I must admit
I admire her for that.

Indulgences, 1883

At my kitchen table,
my bills and accounts before me
(a bushel of coal, a pound of beef, each an hour's work;
a sturdy pair of shoes, at $5, nearly half a week's work;
rent for these three rooms, nearly half my income)
I compute how I am spending my "fortune":

Money saved and sent to Dundee
to have a portrait of my boy made and sent to me
as soon as he will sit still enough for it

A Sunday evening excursion to the theatre in Boston
to finally see the comic opera upon which "The Observatory Pinafore" was
    based
(though it has yet to be performed, I understand)
and to learn the song that now occupies my timid voice
on the walk up the hill in the mornings

A carriage called to take me home from the Observatory
when a cold driving rain would have ruined my skirts and shoes

Sugar for a batch of fudge cut into wee squares
for a morsel of joy at the end of each day

Aye, even with an income such as mine
the luxury of the simplest things

can make one feel as wealthy
as a titan of industry,
as rich as Mr. Andrew Carnegie himself.

An Accounting, 1884

Because I keep some of the Observatory's accounts,
I am perhaps aware of the problem before they are:
the last grant of funds has run out
and five of the male assistants will have to be let go.
They have been at the telescopes at night,
manipulating them into place,
fitting them with photographic plates,
noting the time, the settings, the atmospheric conditions.
They have worked in the darkroom,
preparing the plates, developing the photographs and enlargements.
And they too have done some of the reductions,
all at 40¢ per hour,
that the women computers have done at 25¢.

"Only temporarily," the Director says.
He has been funding some salaries from his own,
certain that more funds will be granted when the photometric work is
    published.
And I have been collecting small sums from the other employees
for some of the equipment with which they do their work,
all to make the funds go farther.

I have wished to suggest that the women,
laboring at least as much as the men,
be paid as the men are.
But it appears that at present
the fact that our wages are low

is all that ensures
that we are paid at all.

An Education, 1884

At the beginning I concerned myself only with the numbers.
Though the equations were new to me, and I was unsure of what they
    meant,
I could rely on my calculations to arrive at the correct reductions to enter
    into the tables.

Then I began to see how the numbers applied
to the brightness, the magnitudes of the stars, the many miles between
    them.
The tables of numbers began to have meaning.

Lately I have studied the previous catalogs of stars—Ptolemy, Herschel,
    and others—
comparing their calculations to ours,
and I begin to understand the reasons for the differences.

I also understand that ours are more accurate.
It is because of the telescopes, the photographs, the spectra, the more
    learned astronomers.

But I cannae deny that it is also because of the women
who are doing the work
of computing.

Invitation to Tea, May 1885

It is suggested that I accompany the Director and Mrs. Pickering
to tea with Mrs. Draper at the Parker House in Boston.
She is reconsidering her husband's legacy of spectral photography of the
stars,
and may perhaps be persuaded to enter into a partnership with the
Observatory.

"Och," I say to Miss Farrar, "I've got naught to wear to such a thing.
And no time, nor funds, to have a dress made now."

How embarrassing then to find that Director Pickering was within
hearing!

"Don't fear, Mrs. Fleming," he interjects.
"Your usual uniform is more than adequate.
We want Mrs. Draper to believe that we would make frugal use of her
funds.
An extravagant costume might give her quite the wrong impression.
It is your work, not your attire, that I hope will interest Mrs. Draper."

So I will go in my "usual uniform"
and try to feel less stung by its "adequate"-ness.
And more confident in the worth of my work.

Stories in the Stars, 1885

The names are familiar from my schoolgirl Latin and the characters of
    myth.
So as I average the estimated magnitudes and calculate the reductions
of the stars within the constellations,
it is not only the numbers that occupy my mind,
but the stories that the ancients saw in the stars:

Carina, Puppis, Vela—
the keel, stern, and sails of Jason's mighty ship Argo,
the one that sailed to find the golden fleece

Andromeda—
the Aethiopian princess chained to a craig in the sea,
saved by the Greek hero Perseus

Pegasus—
the winged horse of Zeus,
galloping across the sky.

Miss Farrar's Discoveries, January 1886

As I prepare manuscripts and tables, I cannae deny my envy
of Miss Farrar, at work on the astral photographs.

Nettie studies the plates, measuring the magnitudes of each star she has
    marked on the glass.
There are 38 stars known to exist in this region of the sky—
38 stars seen and recorded by ancient astronomers with only their eyes,
by modern amateurs with their eyes at small telescopes,
by contemporary astronomers with eyes aided by large refractors.

Nettie, looking only at sensitive glass plates that, affixed to the astrograph,
have received and reacted to the starlight through its lenses,
has found not 38 stars, but 117.

Now she will turn her attention to the spectra of the Pleiades.
What wonders will she find there, among the Seven Sisters?

Waiting, 1886

I have saved the funds for my son's passage, and for my mother to
  accompany him.
My Edward is now nearly six,
old enough to begin school here, to join me in America.

But Mother writes that my grandmother is ill,
that she cannot leave her.
It will be weeks, she says—months, perhaps.

She sends a photograph printed on heavy paper.
My lad has a keen and serious look.
He has my eyes.
He has my dark hair.

I only wish he had me, his mother.

A Romance, 1886

The women computers gather at my home on a Saturday evening.
Miss Bond has brought a magic lantern and will put on a show.

I have borrowed some chairs, candied walnuts and fudge, made tea and hot
  cocoa.
The others have brought biscuits and cake.

We tack a white sheet to the wall,
and in the darkened room, lit only by the pictures projected upon it,
we "travel" to Venice, Cairo, Vienna, London.

The greatest "amusement" to all, however,
is that Miss Farrar has brought along her "beau,"
the dashing Mr. Harris, a publisher friend of her family, visiting from
  Texas.

Nettie is twenty-four, and I am now thirty.
Scarcely six years between us.
But watching her, giddy with romance,
I'm afraid I feel decades older.

I had that once,
or thought I did.
Perhaps I should be jealous, wistful for love.
But I confess (except for the coldest, loneliest nights)
I am not.

I have a son who needs me,
work that occupies me,
stars that provide all the romance I need.

The Henry Draper Memorial, 1886

Resting my eyes and my arm, both tired from use this morning
in the inspection and marking of ink upon the glass plates,
I take a moment outside and find the Director and Mrs. Pickering escorting
    Mrs. Draper around the gardens.

As the result, perhaps, of the tea at the Parker House and the Pickerings'
    persuasion,
there will be a new domed building here,
with the late Mr. Draper's telescope mounted inside,
funds for many more photographs of stellar spectra,
funds for more assistants to handle the telescopes,
funds for more women computers
to calculate the reductions and examine the spectral lines.

Mrs. Draper's investment, I hear the Director exclaim exuberantly,
will be truly astronomical!

Otherwise Engaged, December 1886

Miss Farrar is thrilled to announce her upcoming wedding
and I am thrilled for her.
(I do not allow my thoughts today to linger on my own marriage and
  abandonment.
I do not often discuss it, and when passing acquaintances assume I am
  widowed,
I do not bother to correct them.)

Nettie has shown me what she has learned of the spectral patterns of the
  stars,
making certain that I can continue her work.
(She will, of course, be leaving us, giving up her work
as most young women do when they marry.)

Now I am truly engaged in the work of studying starlight
and I, too, for an entirely different reason,
am thrilled.

Sorrow and Joy, 1887

My mother writes that my dear grandmother, my "Grannie," has passed
away.

When I left for America almost six years ago,
I had not imagined it to be our last parting,
although I knew that age and illness choose their own courses.

When I was a lass
I was found to have a condition of the heart
that might, the doctor said, have caused me to die at any moment.
I spent most of my tenth year in the hospital at Woodside,
and my Grannie brought me books there, and paper dolls and candies, and
taught me to sew,
until I regained my health and was allowed to return home.

When I was no longer a child, but a wistful young woman,
gone swoony over some lad at school,
Grannie told me the story of her husband's mother,
who eloped with a captain of the Highlanders
and gave birth to a son, my grandfather, on the same battlefield where her
husband fell.

I am thinking of her kindness to a young girl,
her compassion for a young woman,
her assistance to a new mother
as I mourn her passing.

Howbeit, Mother says she now has no reason to stay in Dundee,
and will make arrangements in the months ahead to dispose of the house,
stay with my sister and her husband through the birth of their next child,
and make preparations for passage to America with my Edward.

I dream tonight of my Grannie running to her dear husband across the
 Highland heather
and of my own bonnie boy sailing toward his mother across the wide blue
 sea.

Portrait, 1887

My long dark hair,
carefully combed and assembled with a multitude of hairpins

the necklace I am fond of,
a simple ribbon studded with small pearls

my best dress,
a bodice of organza, extravagantly beaded and gathered,
a neckline, though modest, exposing my throat

and at the photographer's insistence,
for contrast against the dark fabric, dark eyes, dark hair,
a corsage of white silk flowers
pinned into place by the photographer's assistant.

He is surprised when I raise my chin and hold my countenance steady
    without instruction,
practiced as I was in the Daguerreotype studio of my father.
And another pose, more demurely downcast.

I will send a print to my mother
so that she can show my son
the mother who awaits him.

More Work, More Workers, 1887

Many of the male assistants,
some of whom went on working without pay when the previous funds
    were spent,
will again be employed at their former salary, $2500 per year.
And there will be more women employed as copyists and computers,
at the salary of $1500 per year.
And recorders, who note the locations and observations of those intently
    examining the photographic plates,
women as well, at 25¢ per hour.

"I leave it to you, Mrs. Fleming," the Director says,
"to determine suitable women to fill our ranks.
I receive letters weekly from those looking for work—
daughters, wives, and sisters of astronomers,
graduates of our few women's colleges,
bluestockings intent on intellectual pursuits,
as well as those in dire straits, hoping for any gainful employment.
I will suggest a few to you, whose connections I am sure of
and who will no doubt be up to the task.
And perhaps," he continues,
"we should consider a dormitory, as some industries provide for their
    women workers.
Let us see what the College will say to that."

I had hoped for the time to examine the plates and review the spectra.
Aye, I'll be happy to have more assistance,

happy to employ more worthy women.

But it is also work to do the hiring and training of new workers.

And a dormitory would no doubt be more trouble than I'd care to
supervise.

When will I get back to the stars?

Form of Records for Harvard College Observatory, 1887

All observations made at the Harvard College Observatory
should be recorded according to the following system.
The size of record books is 20 x 25 cm. . . .
each book contains 216 pages
each is ruled with 31 lines.
Each day's work should begin on a fresh page,
the date being placed at the top
without underlining the date.

The work—
the plates of stars and spectra to be read, the reductions to be made—
has multiplied,
and so have the ranks of women working.

All original records should be entered in pencil, and
no erasures should be made under any circumstances.
If it is desired to change a word or figure,
draw a line through it,
taking care not to render it illegible,
and enter the correction between the lines or following the original entry.
Any subsequent remarks or corrections may be added, in ink,
provided they do not obscure the original pencil record.

There is little time to train our new members.
The volume of work demands that they simply begin it.

The time of every observation should be entered
in the left-hand margin of the record-book in standard time,
twelve hours being added between midnight and noon.
The day is regarded as beginning at noon,
so that an observation made in the forenoon belongs to the previous day.
All records,
especially descriptions of apparatus or experiments,
should be made sufficiently full to be intelligible without further explanation
to any person familiar with the subject, and not familiar with the particular
    research.
when a record-book is not at hand,
notes may be made on paper
and afterwards pasted into the book.

It is,
and I am, I suppose,
meticulous and demanding.
But records of the skies of today
to benefit the discoveries of the future
will be the result.

In copying or computing,
accuracy is of the first importance,
rapidity second,
and the character of the writing, as long as it is legible, third.
Errors should not be erased,
but the correct value interlined,
both to save time and to permit a subsequent recovery of the original
    computation,
which may (upon further review) *prove to be correct.*

The Colors of Light, 1887

It was Fraunhofer, a lens- and telescope-maker, who in 1814
first saw the lines in the rainbow of colors of sunlight.

It was Kirchhoff and Bunsen, physicist and chemist, who in 1859
studied those lines in the flames of evaporating elements.

It was Huggins, an astronomer, who throughout the 1860s
found those lines in the spectra of the stars.

It was Secchi, a Roman Catholic priest and astronomer, who in 1866
divided the patterns of those lines into classes of stars numbered I, II, III,
    IV.

And it was Anna Draper and her late husband, a medical doctor and
    amateur astronomer, who most recently
attached a spectroscope to his telescope and photographed those lines as
    they caught the light of the stars.

It is I, former maid and copyist, in this year of 1887,
who is tasked by Director Pickering with classifying those many patterns
    of lines
that might help to reveal the secrets of the stars.

Och, I only hope I am up to the task.

Emigrants, 1887

Mother writes that she has secured passage
on the ship "The Prussian" out of Glasgow on the tenth of September.

She will bring with her
my sister Mary's boy, already eleven, to live with my brother Robert and his
    family in Boston;
my sister Johanna's little girl, to deliver to her mother;
and my son, Edward Charles Pickering Fleming,
nearly eight, to bring him home to me.

With fair seas and good fortune, they will arrive before October
to join the families that have endeavoured to prepare a home for them in
    America.

I have been waiting so long.

Preparations, 1887

My brother has been inquiring at the port,
and sends word to the Observatory that "The Prussian" will arrive
   tomorrow.

I ask Director Pickering for the afternoon,
and he gives me the entire day, so long as I agree to call at Alvan Clark &
   Sons about a lens.

As there is a small toyshop near Clark's, where I hope to find a
   bilboquet—a wooden ball tethered to a cup,
I am only too happy to comply.
At my little home I have ready
books and paper, pencils and ink, a box of watercolor paints,
and a parade of animals I have cut from a magazine and pinned to the wall
above the small bed my Edward will use.

I arrange for a carriage for the morning.
Och, I cannae wait for the reunion!

My Bonnie Lad, 30 September 1887

I fear I will not recognize him.

The last portrait sent from Dundee was nearly a year ago,
and young boys grow so quickly.

Aye, but I know him.

He is jostling with his cousin, struggling with the bag he is carrying for his
    grannie.
She is close behind with little Johanna in her arms.

There is a flurry of hugs and kisses.
Robert takes the bags, Mother takes his young Andrew by the elbow,
Johanna takes her wee girl into her arms, and—
my heart soars just to say it—
my bonnie boy shyly takes my hand.

Connected, 1887

My love for my son these six years past
has been like a telegraph message—
necessarily disconnected, abbreviated—
a series of dots and dashes sent across the miles.

Now, with him here beside me,
for every smile and giggling laugh,
every frown and schoolboy tear,
my hand brushing his hair from his brow,
his hand tugging my sleeve,
the dots and dashes have connected themselves
into a lovely, flowing script.

Echoes, 1887

I watch him
when he's busy with his sums—
which I can see already he prefers to writing—
and there is something in the way he holds his pencil and looks up and to
    the right
that is a trace of James.

His coloring, dark eyes and hair, proves him a Stevens.
His laugh is like his uncle Charlie's
and something I cannae define, passing over him at times,
recalls my father as I remember him.

What of me, I wonder,
will echo in my son?

Classifying, 1887

I gather from the plate stacks dozens of spectral photographs,
assemble the journals in which I've described the spectra of hundreds of
    stars.
The Director reminds me that he expects a classification system of use to
    astronomers all over the world
as the discovery of new stars accelerates.

However will I organize them all?

Perhaps in the future we will know
their temperatures, their ages, their motions, their compositions.

But now I have only the tiny lines made by each star upon its plate.
So I allow the lines themselves to guide me.

Arabic numerals imply an order I cannot be sure of;
Secchi's Roman numerals are too cumbersome and take up too much space
    in the tables.
So I will use the letters of the alphabet,
But I will leave out "J," it being too like "I" in the German language
in which so many European astronomers work.

So.
A: strong, broad lines only
B: strong, broad lines with other dark lines
C: strong, broad lines in double
D: bright bands appearing in addition to dark ones

E-M, excepting J: the complexity increases
N, O, P: particular spectra of uncommon stars and nebulae
Q: most peculiar spectra. Mysteries.

Aye. Mysteries.

His Eighth, 1887

"The lads at school have marbles," Edward says.
"Could I have a sack o'those?"

I have promised him a toy
and cake with eight candles
for his birthday.

My students at Broughty Ferry played with marbles. I should have
　　remembered that.
And a warm winter coat. He'll need one soon.
And perhaps one of the male assistants will tell me where I might find a
　　football.

Birthday Party, October 1887

Exhausted after the day of work at the Observatory,
preparations for supper, decorating the cake, arranging the table and
    borrowed chairs to accommodate our guests—
Mr. Solon Bailey, the astronomer, and his wife and young son,
my mother and my older brother and his family
(his wife now carrying a bairn to be their third, his daughter a darling of six,
his son a fine companion for my Edward, as well as their cousin Andrew)—
the games of bilboquet and jackstraws, pin-the-tail and blindman's-buff,
the latest news from Dundee,
the little gifts and thanks and goodbyes said,
the sweeping and the dishes done,
tucking my Edward into his bed,
I stay awake just long enough to fall into my own.

In the morning, then, the fire is out,
and breakfast is only slices of apple and a spoonful of sugar stirred into cold
    porridge
before we leave for church.

Along the way I point to the sky and remind my son
that the stars are there, beyond the clouds, beyond the blue,
hidden from us by the sunlight all around us.

but he has found a copper penny on the ground,
and stoops to study
every glint of glass or metal shining from the macadam.

In church he is good, listening and singing hymns,
but when I attend to the sermon
he takes three new marbles from his pocket,
bows his head and pretends to pray
as he rolls them over and over each other in his palm.

Enter Miss Maury, 1888

We are joined in the Computing Room by a remarkable young woman,
   Miss Antonia Maury.

She is the niece of the late Henry Draper, and according to Mrs. Draper,
she helped him in his chemistry laboratory at the age of four,
read Virgil in the original Latin at nine.

She is one of the very few women to have been allowed to attend, and
   graduate, from college, at Vassar.
She earned honors in philosophy, physics, astronomy.

If she were a young man,
there would be a department to lead, a professorship, a laboratory, an
   observatory.

She has only a room in her family's home,
a balcony with a telescope,
a brilliant and restless mind.

She will work here, another woman computer under my supervision.
I expect that she will do excellent work in cataloguing the stars.

I do not expect that she will be satisfied.

Uncle Charlie, 1888

My brother Charles will soon set out for America,
having finished his studies in Scotland.
He'll stay here in our cozy little house on Upland Road
until he gets his feet under him.

We'll be glad to have him.
Edward loves his uncle,
and without a father here
it will be good for him to have a man about.

But I will have a word with my clever young brother
about the drinking song (about a Highlands lass and a bagpiper, as near as
    I could make out)
that he taught my son back in Dundee.
I'm sure Charles thought it quite a lark
to tease his sister through her innocent son!

Charles has always been a tinkerer,
a bit like our father was, though he hardly remembers him.
In Boston he intends to be a watchmaker.
And we have a clock needs fixing,
a doorbell that will not turn,
a lock that no longer responds to its key.
I will have him set his clever mind to that.

Binary Stars, 1889

It has been the subject of much speculation between Miss Maury and the
    Director
since they noticed the evidence on a series of photographic plates:
What to make of a star
whose spectra changes and changes back again?

Now Miss Maury asks a different question:
What if they are watching not one star, but two?

Two stars revolving around each other—
first one the nearer, then the other?

They find they can predict the pattern,
time the intervals of the stars circling each other
at hundreds of miles a second.
Director Pickering identifies the first pair.
Miss Maury identifies the second.

"Do you see, Mrs. Fleming?" Miss Maury asks.

Och. Yes. I see.
And now that I know what I am seeing,
I know I have seen this before.

I go back to the plates that I have studied, labeled, computed, categorized.
By the time the week is out

I have discovered
the third "spectra-scopic binary" star.
And I am quite sure I will discover many more.

Another Heiress, 1889

"A telescope with a lens three times this size can gather nine times the light!"
  Director Pickering says.
Imagine the faint stars, now beyond our vision,
that such an instrument would allow us to see!

He has found another heiress, Miss Bruce of Manhattan.
She has no scientific expertise, only curiosity and generosity and a fortune
  to disburse.

The Bruce telescope will cost upwards of $50,000.

I calculate that, with the stroke of a pen in her cheque book,
Miss Bruce advances Science
with an amount equal to more than thirty years of work at my salary.

What would I do, I wonder, with such a fortune?

How I Wonder What You Are, 1889

In a bit of a reversal, I suppose, from the song he had learned from his uncle, Edward has taught Charles all of the verses of the song he learned at school so they can perform it tonight after dinner:

"Twinkle, twinkle, little star,
How I wonder what you are!
Up above the world so high,
Like a diamond in the sky.

When the blazing sun is gone,
When he nothing shines upon,
Then you show your little light,
Twinkle, twinkle, all the night.

Then the trav'ller in the dark
Thanks you for your tiny spark,
He cannae see which way to go,
If you dinnae twinkle so.

In the dark blue sky you keep,
Often through my curtains peek,
for you never shut your eye,
'Til the sun is in the sky.

'Tis your bright and tiny spark,
Lights the trav'ller in the dark,
Though I know not what you are,
Twinkle, twinkle, little star!

"Bravo!" I applaud, as they each make an exaggerated bow.

"Well, Mina," Charles says, "what is the answer?
What are the stars made of, after all?"

I set down my mending and pour the tea.
"Well," I begin, "it is assumed that the stars are made
of the same stuff as the Earth. Silica. Iron. Carbon."

Edward reaches for the sugar. "Sand, and stones, and dirt," he says.

"Aye," I say. "Sand and stones and dirt, I suppose.
But this is what your mother is helping to discover.
Someday, perhaps, we'll know for certain.
No, not 'perhaps.' Most certainly.
You, my dear," I say to Edward, "will know much more about the stars
than I will ever know."

"Because you're such a clever lad," Charles contributes.

"And because Mam is an a-stron-o-mer," Edward says carefully.

"Well, nearly one," I say.

I cannae claim it is only the tea that warms my heart this evening.

Southern Sky, 1889

I have never with my own eyes seen the night sky of the Southern
    Hemisphere,
but now I can see each Southern constellation in the light it impresses on
    the photographic plates.

They come carefully wrapped from Arequipa,
on the backs of donkeys down the mountain—an extinct volcano, they
    say—
overland into the Peruvian plaza,
onto oxcarts or barges to the port
and a steamship for weeks on the sea
until they reach the port of Boston or New York,
the crates loaded again onto wagons and carted again to Cambridge
where the precious panes of glass are unwrapped, numbered, catalogued,
and finally, in our Computing room at the Observatory in Cambridge,
the light of the Southern stars reaches us.

My Work, 1890

Years

of careful study observation, computation, classification:

I have assembled into tables

the stars,

assigned them each a number,

noted their earlier designations in the Harvard catalogue, the German, the
   Argentine,

their locations in the map of the sky,

the photographic plates on which their images reside,

their magnitudes,

their spectral classes,

the details of their measurement,

remarks of error and speculation,

their constellations,

their elegant Greek letters.

Now

that I see them all assembled, I count the stars that I have catalogued.

They number 10,351.

Recognition, 1890

The names on the title page are
Draper, Pickering, Bache, Harvard
and also the printer: *John Wilson and Son.*
But *The Draper Catalogue of Stellar Spectra* is the work of many more
and for the first time,
my name is among them:
the greater portion of this work, the measurement and classification of all the
spectra, and the preparation of the Catalogue for publication, has been in
charge of Mrs. M. Fleming

It is noted only on the second page of the preface
of the 400 pages of the catalogue,
but I am proud to see it there.

A Raise, 1890

The publication of the Draper Memorial Catalogue, Director Pickering
    says, is quite an achievement,
worthy of a raise in pay.

I still earn less than most of the male assistants, who,
I must note,
arrive later, leave earlier, and always seem to have more time for their own
    pursuits than I.

Yet this will mean the world to me and my son.
I have found a woman who, for the same amount as this pay increase,
will work a few hours each day cooking an evening meal for us, and
    sweeping or dusting or scrubbing some part of our home
with the time that is left.

Och, it means only a hot meal for us, when I've worked until dark,
and someone there when Edward walks home from school,
and the dust regularly shaken from the rugs.

But to the woman, Marie, it is enough to get by.

And to me, it is enough for a moment's peace
after I've finish one kind of work
before I begin the other.

On the Plates from Peru, 1890

I have been occupied with hiring new computers,
documenting the progress of the brick building,
inquiring about the lenses for the new Bruce telescope,
preparing manuscripts for printing,
arranging for supplies to be sent to the observatory in Peru, where Harvard
    astronomers are photographing the stars of the Southern Hemisphere.

But whenever I can, I study the plates.

I have found a new variable star in Delphinus and the binary Beta Lyrae.
And on another plate from Peru,
I note a peculiar haze with a notch—
a semicircular indentation
5 minutes in diameter
30 minutes south of Zeta Orionis.

Whatever, I wonder, might that be?

Exit Miss Maury, 1892

Her ideas take time.

She is finding, in the spectra of the stars,
more detail than my classification system allows for.
She intends to create her own.

She chafes against the tasks assigned to her.
She bristles at giving over her work to anyone else.

She argues—stiffly, formally, politely—with the Director.

She has decided to leave us, at least for a while.

Brick Building, 1892

Work has begun on a new building
to house the ever-expanding collection of photographic plates of the stars.
It will be made of brick—safe from the fire that is the stuff of my
  nightmares.
Mrs. Draper, too, has worried about flames destroying her husband's
  legacy.

I have been derided, of course, by the computers and young assistants,
for the precautions I insist they follow,
for prohibiting their smoking pipes in the computing rooms,
for the soda-acid extinguishers I have placed about the Observatory,
for the occasional drills I impose.

But we all have been touched by the danger—
homes of friends or relations destroyed;
the elegant Manhattan theatre, owned by the Drapers, demolished;
entire blocks of cities turned to ash
by combusting chemicals, a gas lamp overturned, a stray cinder on carpet
  over a wooden floor.

Let them mock my seriousness.
I must do what I can to protect the records of the skies.

The Transit of the Plates, 2 March 1893

There are stairs—too many stairs,
a small expanse of lawn,
and a shallow ditch with a wooden plank across it
between the storage rooms in the Observatory
and the new accommodations in the Brick Building.

I have seen to the packing of 300 crates,
100 glass photographic plates in each.
While I cannot bear to think about the laborers carrying the heavy crates
down the narrow stairwell, across the muddy ditch,
dropping, falling, breaking—
neither can I bear to look away
as they proceed with a different plan:
pulleys and cables
from a roof of the Observatory to a window in the Brick Building,
the crates swinging along in the air
like damp laundry on a tenement clothesline.

The fragile plates all survive.
They have arrived in their—in our—new home.

A Lesson from the Dolphin, 1893

My discovery of a variable star in Delphinus
has drawn the attention of two notable astronomers.

These men have determined that I am in error.

My star, they say, is not variable.
They are sure of their observations, made by each
with their eyes to their telescopes, unaided by astro-photography.

They meet to discuss the source of my error,
compare their notes,
look to the dolphin to pinpoint the star in question
and discover
that they have been looking
at two different stars.

Neither, they discover, is <u>my</u> star.
If they had consulted the astro-photographs,
they would not have committed <u>their</u> mistake.

I am not, perhaps, yet, a notable astronomer.
But I am a woman in command
of a vast collection of photographs of the stars.
The men may not deign to listen to me,
but the stars speak for themselves.
In confronting these men's error,

the photographs of the stars
have spoken for me.

A World's Fair, 1893

Mrs. Draper has been to Chicago, to the Columbian Exposition,
a White City with a tower of electric lights,
a giant wheel that takes fair-goers for a ride in the sky,
a sculpture of the Venus de Milo—in chocolate!
And an entire building, designed by a woman architect,
of women's art and accomplishments.

I had hoped to go as well, though I've not been more than a few miles west
    of Cambridge.
The Congress of Astronomy and Astro-Physics will convene there,
and I have prepared a display of photographs of our work here
and ourselves (crowded into one end of the Computing Room, the women
    at their light stands and magnifiers).
I have been invited to deliver an address on "A Field for Woman's Work in
    Astronomy."

But I must supervise the completion of the move to the new Brick
    Building,
the mounting for the Bruce telescope,
the small army of computers and assistants.
And I would not want to have Edward miss the beginning of school,
nor be left alone,
with Charles now in his own home with his new wife, Eliza.

I shall send my words to be read in Chicago.
I have too much work to do here.

Insult, 1893

"You have been insulted, Mrs. Fleming," Director Pickering fumes.
"Mr. Chandler questions our observations.
His new *Catalogue of Variable Stars* omits your recent discoveries.
He denies them only because he cannot see them himself
with his eye to a telescope!"

Perhaps.
But I wonder, if Mr. Chandler—who worked here briefly
and knows that "M. Fleming" is a woman,
still a rarity in the annals of astronomy—
would question the veracity of my discoveries
if he were unaware of my background or my sex.

"Alleged but unconfirmed," Chandler's notes read about a dozen of my
    stars.

"Alleged but unconfirmed," is it?
Och,
it is Mr. Chandler's discoveries,
with only his eye and his notes to confirm them,
that are suspect.

I have the photographs.
I have the evidence.
It is here, in the plate collection, for all the world to see.

"A Field for Woman's Work in Astronomy," 1893

In the earliest records of ancient Greek History
we can trace the great interest
which centres in the heavenly bodies,
and in Astronomy, the greatest of all sciences,
but in no way do we find women connected with the study of this science
until a comparatively recent date.
Caroline Herschel, Mary Somerville and Maria Mitchell
were, as women, pioneers in this work.
We can not say these were the only women of their time
capable of devoting themselves successfully to this work
and of adding to our knowledge
of the heavenly bodies and of the laws which govern them.

A great many women of to-day
must have a similar aptitude and taste for Astronomy—
and if granted similar opportunities
would undoubtedly devote themselves to the work
with the same untiring zeal,
and thus greatly increase our knowledge
of the constitution and distribution of the stars.

The United States of America is a large country,
with a largehearted and liberal-minded people.
Here they have made room for comers from all other countries,
have welcomed them and have given them a fair open field
and equal advantages in pursuing their labors or studies.
There is no other country in the world in which women,

not as individuals, but as a class,
have advanced so rapidly as in America,
and there is no other country
in which they enjoy the same unlimited freedom of action
which affords them the opportunity to find their own level.
In their studies they encounter very little narrow-mindedness or jealousy
in their brother students or fellow workers in the same field of research,
but in general they are treated with the greatest courtesy,
encouragement and assistance being graciously accorded.

Women, therefore,
who have taken up any branch of science,
or indeed work of any kind,
need not be discouraged in it
even if one or two
of the great mass which goes to make up the whole
in their superior judgment
refuse to give credit to their work.

While we cannot maintain that in everything woman is man's equal,
yet in many things her patience, perseverance, and method
make her his superior.
Therefore, let us hope that in Astronomy,
which now affords a large field
for woman's work and skill,
she may, as has been the case in several other sciences,
at least prove herself his equal.

Labor honestly,
conscientiously,
and steadily,
and recognition
and success
must crown your efforts
in the end.

Letters, 1893

The Director's own handwriting is nearly illegible,
even, sometimes, to himself.
There is in his office a new type-writing machine,
but for the Observatory's business,
only the more formal pen and ink is appropriate.
So at some time in nearly every work day,
he and I confer to address the Observatory's correspondence,
and I later—when my sore arm is up to the task—take pen in hand.

There are letters of introduction and reference
from women looking for work.
They have heard that the Observatory hires women—
one of the few workplaces, outside of a factory or a laundry, that does.
Some are daughters, sisters, wives, widows of scientists
steeped in astronomy or mathematics for much of their lives.
Some are graduates of women's colleges,
or autodidacts, well-studied in their own libraries.
Some are orphans, widows, abandoned women
in dire straits that I understand all too well.
I keep a file of possibilities for whenever there might be an opening.

There are letters of inquiry and discovery
from amateur astronomers and curious observers.
They have seen a meteor as they walked home at night,
or have studied the same constellation for years

and now have seen something amiss.
Some respond to the Director's call for careful observations from the
    public.
I record their reports, review the historical record,
find the photographic plates that match their notes,
and reply with the results of the investigation.

There are letters of invitation and collegiality
from astronomers who wish to visit here,
or to welcome Harvard astronomers to colloquia or college lectures.
I keep a calendar,
send the acceptances and regrets,
make the arrangements.

There are letters of information and solicitation
to potential partners and donors,
subscribers who contribute to the Observatory's finances
in exchange for periodic postings of new developments.
I send the Observatory's thanks,
request their continued support,
send them our news of equipment, facilities, and discoveries.

There are more delicate matters,
letters to the Director's brother, Professor William Pickering,
who has been using the Peruvian telescope to observe the planets,
claiming to see lakes and canals on Mars,
implying the existence of some race of Mars men.
I write the Director's request that he return to the agreed-upon work,
focus on the stars,
and refrain from embarrassing the Observatory.
There are letters from Miss Bruce and Mrs. Draper,
supervising the use of the funds they generously contribute.
I send an accounting along with profuse thanks,
emphasize the advances dependent on their continued support.
And the conflicting letters regarding Miss Maury:
from her father Reverend Maury,

requesting patience with her mental state and apparent physical
    exhaustion;
and from her aunt Mrs. Draper,
requesting the Director hurry her to finish her classifications and publish
    her work;
and from Miss Maury herself,
concerned that in her absence her work will be overtaken, denying her due
    credit.
For each of these I assist in finding the most diplomatic words
and write the Director's assurances and support.

I write a copy of each letter for our files.
The letters are signed by the Director,
the replies addressed to him.
I am occasionally complimented on my penmanship.

Thousands of Stars, 1893

I receive letters from Mrs. Bailey in Peru.
Mr. Bailey has taken charge of the Observatory there, and she assists him.

The clarity of the skies, she says, is breathtaking.

They have trained the new telescope on nebulae, and now believe them to
consist of thousands of stars.

"Look there, Mina," Ruth Bailey writes, "in the globular clusters. It is as if
a chest of jewels has been opened."

The Baileys are diligent.
Entranced with the view through their telescope,
they nevertheless maintain the schedule of methodically photographing
the skies and sending the plates to Harvard.

When I write to Ruth again,
I am happy to tell her that I have found a new variable star in Omega
Centauri.
And there are sure to be many more.

Nova Normae, 26 October 1893

I expect nothing remarkable when I select a plate newly arrived from
   Arequipa.
The plate contains the spectra of stars in Norma,
a four-sided constellation named not for a goddess, but for a carpenter's
   tool.

But here, here is a unique spectra—a dozen bright hydrogen lines,
more and brighter than I have seen before—
though I know this is not the first time
that I have studied this part of the sky.

I find Director Pickering, who takes my place at the light lectern
while I search the collection for earlier plates of this location.
I find not a trace of it upon the plates,
though other, dimmer stars are easily visible there.

The Director snaps his fingers. "Do we have the photograph of Aurigae?"

The star was discovered just a year ago,
by a Scottish amateur with a simple telescope.
Fortunately, he alerted professionals who managed to photograph its
   spectra
before it faded.

We do have Aurigae.
And it is so like my star as to be its twin.

The spectra is that of a nova—a new star bursting to life,
or, perhaps,
hurtling toward its death.

The Director cables Mr. Bailey in Peru.
If he can again focus his telescope on the star,
he may catch it as it shrinks and fades.

There have been only nine such new stars recorded in the history of
    astronomy.

This, Nova Normae, is the tenth nova ever to be discovered,

the first to be discovered by spectral photography,

the first nova to be discovered—
as Director Pickering announces to the Astronomical community and the
    press—
by Mrs. M. Fleming.

The Bruce Telescope, 1893

Miss Bruce's money is well invested in the new telescope.

The iron frame that will hold the instrument is installed today,
an entire day requiring six strong men and four horses
to haul and hoist and assemble.

The lens itself is two feet across,
ninety-pound discs of perfect transparent glass,
ground and polished and assembled together
to direct the light upon the photographic plates.

In Miss Bruce's wish she quotes a philosopher:
that with her gift we will explore the "luminiferous ether."

Where Do the Years Go? 1893

I was fourteen years of age
when I took a position teaching younger children
at the Broughty Ferry school back in Dundee.
I admit I was proud to help support my siblings—young Johanna and
    Charles—
my widowed mother,
my grandmother,
my grandfather still alive then, as well.

Now my own son is fourteen.
I admit to some pride
that he does not need to do as I did.

There is no need for him to abandon his studies,
to go to work
to keep food on our table,
a roof over our heads,
though I keep a cautious eye on my salary and our expenses.

For his birthday gift
I have asked the bookseller
for a collection of Poe's mysteries with the detective Dupin,
which I have resisted until now, Poe being so sensational.
(I suspect Edward has read them once already,
borrowed from a rougher friend,
though he would not tell me so.
It is time to allow him to make his own choices in life.)

Och,
but I wish sometimes he would realize
what a lucky lad he is!

Harvard versus Penn, 30 November 1893

I want Edward to stay warm,
call out to him to remember his gloves,
but he is out the door,
his football under his arm
and a school chum at his side.
Ah, well.
He has pockets to keep his hands warm,
though they are usually filled with interesting stones, castoff horseshoe
    nails, thick, smoothed-edge pieces of broken green glass.

I fold a wool blanket over my arm.
I've a knit cap and gloves
and a pair of Edward's socks over my wool stockings
so I'll not catch a chill while I watch the Harvard Eleven—
undefeated so far this year—play today.

I had hoped to sit with Edward and direct his attention
toward the remarkable young law student playing for the team,
by all accounts the first Black man to play football here.
His parents, according to the newspaper, were formerly enslaved.
Yet William Henry Lewis is here, at center rush,
at least twenty-five pounds slighter than his corresponding opponent,
using wit and strategy and strength to hold his own in the game.

"Do you see?" I would have told my son.
"He is not the largest, nor had the most advantages,

but he has used his mind and determination
to make his way in the world."

But I have no chance today for this conversation.
Edward finds a place with his friends, while I find my own.
Perhaps he is already
making his own way in the world.

(Harvard wins, 26-4)

An Ache, 1893

The fatigue is familiar.

There is the strain on the eyes
from the close looking at the glass plates, at the scribbled notes,
at the columns and columns of tables in the star catalogues.

There is the heaviness in the legs,
from the feet tucked under a desk most of the day,
toes pressed into the tips of their shoes,
the walk home down the hill and the work to be done there then.

There is the ache in the arm and shoulder,
from lifting the glass plates,
grasping the heavy magnifying glass,
the pages and pages of writing.

I can remove my shoes at the end of the day,
soak my feet in a basin
or prop them in a chair.

I can rub and rest my eyes,
shield them against the light,
close them in the dark.

But though I stretch and shake my hand,
rest my arm, rub my shoulder,
I cannot avoid the use of them.

Och,
I am not a laundress, a coal miner, a peasant farmer,
working my fingers to the bone,
scorching my hands,
dirtying my lungs,
breaking my back in the fields.
I have the privilege of work that uses my mind more than my body.

Still, the pain in my arm and shoulder grows.

Enter and Exit Miss Maury, 1894

She has come and gone again.
She has—quite late—submitted the work that concerned her:
her new system for classifying the "Spectra of Bright Stars."
The Director has promised her name on the title page as the author.

It needs only, Director Pickering tells me,
to be edited and prepared for press.
That task falls to me.

At times it seems
I cannae bear
another responsibility
falling to me.

Misses Leavitt and Cannon, 1897

They cannot be admitted, at Harvard, being women.
They cannot attend classes here, cannot earn degrees here.

The only way they can learn, at Harvard, is to work here.

Director Pickering has invited Miss Henrietta Leavitt and Miss Annie
    Jump Cannon,
students at Radcliffe and Wellesley women's colleges,
to accept unpaid assistantships
so that they may study the stars here
and advance the aims of the Observatory.

Och.
They must come from families of means, if they are able to work without
    pay.
They must be very capable women, to be invited by the Director.

Miss Leavitt will study the northern stars,
assessing their magnitudes from our photographic plates.
Miss Cannon will study the stars, as well,
but she will be the first woman here to hold another honor:
she will be allowed to use the telescope.

An Educated Man, 1897

I will confess that I once hoped it would be Harvard that welcomed my son
when he finished preparatory school,
to have him near me, to hear news of his successes, to be warned of his
    difficulties.

But my Edward struggled with his Latin grammar,
avoided writing his essays,
remained uninterested in Law or Medicine.
He has worked in the computing room for a month at a time,
assisting capably with the reductions,
speaking knowledgably enough with the computers and assistants,
politely attending the Observatory's events with me.
But he is not inclined toward Astronomy,
tending more toward the Earth than toward the skies.

Today he begins his studies at the Massachusetts Institute of Technology,
in Mining, Engineering, and Metallurgy
(explaining, I suppose, the trouser pockets always full
of found coins and stones and metal bits,
his eyes downcast not due to disposition
but to interest in what they might find along the ground).

I have scrimped and saved, and must continue so, to fund his studies.
He will board at home to avoid the expense,
with a very small allowance so he will better learn the value of money,
and take the ferry, or electric train car, or carriage

to Boston and Copley Square.
(Perhaps he will try the new subway car—something I have resolved to do.)

I have had him to the tailor
(Johanna's husband James Mackie, who gives me a good price) for a suit
   and hat,
to my brothers John and Charles for slide rule and compass and magnifier,
to the stationers' for journals, loose paper, pens and ink and pencils,
though he insists he has more than enough.

I could never provide the advantages of
a high birth, a luxurious life, a father.

So I must supply what small encouragements I can
to make sure my lad becomes an educated man.

A Request, 1897

I request a moment to discuss the problem with the Director.
We sit in his office, at his clever revolving desk,
photographic prints, maps, manuscripts, letters between us.
The problem, I tell him, is the delay
in receiving photographic plates from Arequipa.

There have been difficulties in South America, of course:
an earthquake which rattled the buildings—
but the telescopes, on heavy solid piers, held fast;
the Peruvian civil war when Mr. Bailey dismantled the telescope,
burying the enormous, heavy lenses for safekeeping
until the danger of marauders had passed.
But the observatory there is now in fine form.
Why are we not receiving the plates in a timely fashion?

The Director removes his glasses, using them as a conductor with a baton.
"It seems, Mrs. Fleming," he says, "the assistants are frustrated.
They man the telescopes. They take the photographs.
They want the credit for the discoveries.
They are now taking the time to look over the photographs, noting any
    outstanding spectra.
They wish to alert us here when they see something significant."

I am indignant.
It is not the credit that concerns me.
We, myself and the women under my supervision, examine the plates.

It is more than noticing something out of the ordinary.
It is computing the magnitudes, measuring the distances, analyzing the
     spectra,
consulting the historical record, examining the appropriate plates,
categorizing, cataloguing, caring for them.
It is time-consuming, painstaking work, tedious at times,
but rewarded occasionally by discovery.
"Please," I beseech the Director, "please do not allow that to be taken from
     us."

He pauses, rotates his elaborate desk one-twelfth of a turn,
picks up and sets down another photograph.
He rubs the bridge of his nose and replaces his spectacles upon it.
"I will speak to Mr. Bailey," he says.

Before the century turns,
I have discovered Nova Aquilae and Nova Sagittaris
on the plates of the Southern sky.

The Amusing Story, 1898

I hear—not for the first time—
the not-entirely-true story of my employment here:
how it was (when I was employed in the Residence, and not yet in the
  Computing Room)
that the Director, frustrated with his male assistants' lack of industry,
cried in exasperation, "My Scotch maid could do better!"

And, eventually, I did.

The small company of astronomers find it witty,
a gibe at the former assistants, some of them now renowned scientists.

I wonder that they do not think of the insult implied—

to a Scot, to a maid, to a woman, any woman, and to me—
of how amusing it is to think that a woman, a maid, a Scot,
could do the job equal to—better than—a man.

What I suspect, though,
is that now that I have applied the formula to the notes that the
  astronomers make each night with their telescopes,
computing the distances between the stars;
now that I have done the work of decoding the striated patterns
of the photographic plates and found that some stars
are not simply single stars but clusters of them;
now that I have guessed at the composition of the very fires

that burn to make each of them visible;
now that my work has earned recognition,
sometimes equal to their own,
now that I've made myself more than useful,
that the story has begun,
the amusing story
(quite different from what I remember)
that they use to explain my success, and my presence—
my working class, immigrant, female presence—
in the company of these men.

The Bruce Medal, 1898

Miss Catherine Bruce, the benefactress of the telescope
now revealing so much in the sky of the Southern Hemisphere,
has another gift to bestow:
an award for an astronomical researcher with a life-long record of
   achievement:
The Catherine Wolfe Bruce Gold Medal of the Astronomical Society of
   the Pacific.

The Director assembles a list of possible recipients
for the first award of 1898,
as Miss Bruce herself is easily exhausted at the age of eighty.
She has, however, opinions.
She wishes to stipulate that the award be not limited to men alone,
nor to Americans alone, but to *"citizens of any country, persons of either sex."*
For the first honor of 1898, however, the Society's Board of Directors
has not been so imaginative as she.
They have chosen the eminent astronomer and mathematician Mr. Simon
   Newcomb,
a worthy choice, I'm sure.

An Astronomical and Astro-physical Society, 1898

The Director has invited every prominent astronomer.
Simon Newcomb is here.
Mr. Hale, of the Yerkes Observatory.
Our own Professor Searle,
Mr. Bailey just returned from Peru,
and dozens more eminent scholars.
Our own computers and assistants round out the participants,
and my own Edward Fleming, the youngest delegate, conveniently
    represents both
the Institute of Technology and our own membership.

I have been planning the gathering with Mrs. Pickering—
the best-situated meeting rooms,
the luncheons and social gatherings,
an outing by railway to Blue Hill to see the Meteorological Station.

We did not, however, plan on the heat wave.
On the first day it is far too stifling in the Observatory,
so Mrs. Pickering graciously opens the residence, recently expanded,
the gathering of over one hundred in her drawing room justifying that
    effort and expense.

I have prepared a short paper on "Stars of the Fifth Type in the Magellanic
    Clouds,"
which the Director is happy to read to the assembly sitting amongst the
    Pickerings' fine appointments.

After detailing the spectra of bright hydrogen lines of the stars we have
     found in the last fourteen years,
Director Pickering looks up and continues to address the group.
"What Mrs. Fleming has failed to include in her paper," he says,
"is that of the seventy-nine newly-described stars,
she herself is the discoverer of nearly every one."

How disarming to hear the applause that erupts from this announcement,
the calls for me to stand and be recognized,
the questions from the learned astronomers—my colleagues—
that I am able to answer!

And how gratifying to remember, as I regain my seat,
that my own son is witness to the presentation!

When the assembly takes to Blue Hill on the final day,
watching the ascent of the kites outfitted with the tools of measuring wind
     and temperature,
I feel that I might reach the very same heights!

A Title, 1899

It does not come as a surprise,
since I have assisted with the related correspondence:
the Director's petitions to the Harvard Board;
the lists of my accomplishments, the duties I fulfill here,
the importance of the store of photographs
to the larger Astronomy community.

Still, the announcement is remarkable:
There is a new office at Harvard University
and I,
Williamina Paton Stevens Fleming,
have received its title:
Curator of Astronomical Photographs.

I am the first woman to hold any title at the Observatory.
I am the first woman to hold any title at Harvard.

Aye, but to me
there is something more important than this:
the position ensures that there will always be someone to care
for the precious collection
of photographs of the stars.

Photo Gallery

Photographic portrait of Williamina Fleming with necklace and corsage, 1880s (Harvard University Archives).

Director E. C. Pickering with Williamina Fleming and Harvard Women Computers in the East Computing Room, 1891 (Harvard University Archives).

Harvard Woman Computer recording as Williamina Fleming studies a glass plate (likely a staged photo), 1891. (Harvard University Archives)

Williamina Fleming, Harvard Women Computers and the crew of the Minia aboard ship, 1910 (Harvard University Archives). Mina is bottom row, center, in dark dress and hat.

Edward Pickering Fleming and M.I.T. 1901 Class Football Team, 1900. Edward is at bottom center with crossed legs. (M.I.T. Technique Archives)

Group Photo of the American Astronomical Society meeting at Yerkes Observatory, 1909: American Astronomical Society. Mina is bottom row, with umbrella. (Hanna Holborn Gray Special Collections Research Center, University of Chicago Library).

Group Photo of American Astronomical and Astrophysical Society at Harvard, 1910 (Harvard University Archives). Mina is lowest row standing, Fifth from right in dark clothes with light brooch.

Williamina Fleming, Curator of Astronomical Photographs at Harvard University, in the Collection, 1909 (Harvard University Archives).

The Turn of the Century,
1899-1900

I have organized a small celebration at the Observatory
at the end of the workday on December 30th, a Saturday—
tea and sandwiches and candies and fudge, a simple gathering—
but with toasts to the Old and to the New, the mood is jovial.

The scientists extol the advancements of this last one hundred years:
the electric light and telegraph,
the photograph and astro-photograph
("Hear, hear!" from the Director to this),
the microphone
(Hear, hear!" from Miss Cannon, who uses an aid to her hearing—
drawing amusement from all at her double usage of "hear"),
the typing machine and sewing machine
(applause from many of the women present),
the telephone and bicycle
(applause from the Director, himself a bicyclist),
the locomotive and railroad
("Choo, choo!" from one of the assistants).

"And what will the next century bring?" Professor Bailey asks.
"An auto-car!" shouts one.
"A flying machine!" shouts another, to laughter.
"A computing machine!" suggests one of the women computers, hopefully,
although I should not like to see a machine take over the work of those
    under my charge.

We bid each other "Happy New Year!"
Some have plans for elaborate celebrations in Boston or Manhattan,
others remaining in Cambridge.

With Edward staying near the Institute with friends and Miss Cannon
    visiting her sister,
I return home to Miss Hegerty who has a meat pie and potatoes in the oven.
We eat a bit and save the rest, plan a list for the grocer for the next week,
and I send her home with a portion and a bottle for tomorrow
and my wishes for her own Happy New Year.

I had planned on church in the morning
and a celebration at my sister Johanna's tomorrow,
but my arm is sore and I am exhausted.
I decide to spend the beginning of the New Year quietly at home,
feeling rather feverish,
and hoping that the new century will bring,
along with the auto-car and flying machine,
new stars for me to discover,
and equal advances to medicine
so that I may remain well enough to discover them.

A Journal, March 1900

The University has determined to produce a record for posterity:
a Chest of 1900,
with accounts and artifacts from life at the turn of the century.

I am requested to keep a journal of my life,
at work and at home,
for the month of March.

Though I write often to friends and family,
recounting my thoughts and activities,
I do not regularly keep a journal.
To me it seems an unnecessary extravagance,
with the only audience myself.
But for Harvard, I will make the effort.

"Journal of Williamina Paton Fleming, Curator of Astronomical Photographs, Harvard College Observatory"

In the Astrophotographic building of the Observatory
12 women, including myself,
are engaged in the care of the photographs;
identification, examination, and measurement of them, reduction of these
    measurements,
and preparation of the results for the printer.

From day to day
my duties at the Observatory
are so nearly alike
that there will be but little to describe
outside ordinary routine work.

My home life is necessarily different
from that of the other officers of the University
since all housekeeping cares rest on me,
in addition to those of providing the means
to meet their expenses.

Och, just the writing of this
takes time away from the work that needs doing!
And though I do not intend it,
the inequity in being a working woman
and concern for my finances

cannot help but become
part of the occupation of my days.

Well, if Harvard wants an honest accounting,
for posterity,
it shall get it.

The Journal: Evenings Out, March 1900

Now that the sunlight stays long enough each day to see me home from the
    Observatory,
I am more apt to venture out after the long day of work.

On Thursday I joined my friends Mrs. Solon Bailey, Miss Anderson, and
    my sister Mrs. Johanna Mackie
at the Castle Square Theatre, a glorious place for a show,
where we sat in the second balcony for "The Firm of Girdlestone,"
which we all enjoyed thoroughly.
Mr. Arthur Conan Doyle's play does not employ his famous
    detective—quite the sensation recently—
but has more in common with the doings at a workplace like ours at the
    Observatory,
than a murder on the moors.

I would have liked to accept Mrs. Bailey's invitation to dinner and
    overnight at her house in Boston,
but I took a carriage home, because my little family—
Edward and his friend Neyle, who has been with us since Christmas,
and Miss Cannon, who is lodging with us to conserve funds
and to remain within walking distance of the Observatory—
is apt to be late for breakfast and the day's duties
if I am not present to get them going in the morning.

On most Sundays, I "make the rounds" of social calls on my friends and
    relations—

to Roxbury to see Mrs. Atherton Brown and her young musician Alice,
to Mt. Monadnock to call on Mrs. Hill and my sweet little friend Edith
    Scott
(still suffering from a cold, poor thing),
and was able to thank her father
for the beautiful volume of Scottish poems he sent me at Christmas,
to call upon Captain Adams of the "Minia," making the crossing again this
    summer,
to my sister (whom I hope to convince to do some work at the Observatory,
when her young Johanna takes her leave),
and often the the growing families of my brothers, the three of them near
    Cambridge,
with my nephews and nieces now aged from grown to toddling,
and my mother dividing her time between them,
staying wherever she is most needed.
I often carry the toys I have found in the shops in Boston
or the dolls I have sewn in their Highland costumes
so the wee ones might play at being Scots
like their fathers and aunts and grannie.

But any Sunday out
must end in time for me to return home by eight o'clock
to meet Miss Christopherson, whom Dr. Smith sends now to massage my
    arm.
Last fall, he was afraid that I might lose the use of it,
but with the physiologist's help, I am still able to do
my sewing, my letter writing, my computing, my work.

The Journal: Evenings In, March 1900

It is a pleasure to arrive home
and find Miss Hegerty with dinner nearly ready to set upon the table,
or Edward and Miss Cannon at work in the kitchen.
Then we sit together for supper,
and make quick work of clearing the dishes, scrubbing the kettles,
and putting it all away.

Then we may play "India" rummy,
enjoy some fudge or hot cocoa,
with Edward's young friends or Miss Cannon's cousin Adelaide Jump
    visiting,
or Johanna or Miss Anderson stopping in to discuss again the play or our
    reading,
the "news" of the day as reported by the Herald,
the doings at the Observatory.

On Saturday evenings we may invite a wider circle of friends, old and
    young,
for games and a "spree" of treats that my little family has helped to prepare
    throughout the week—
pitted dates with peanuts, smoked oysters, creamed walnuts, cakes and
    sweets—
the guests revolving around the tables like variable stars,
playing jackstraws, then crokinole, then cue ring,
changing partners in lively fun and conversation,
the evening not ending until those remaining

have talked ourselves almost to sleep,
falling to bed exhausted half past midnight.

And on other evenings I am deserted,
the lodestone left to keep the house from blowing away
while Edward goes to study with his friends
and Miss Cannon slips away to the Observatory if the skies are clear
(though I wish she would more often work on those remarks for
    publication).
Then I sit by the stove, attempting to stay warm
while mending clothes or sewing for charity—
doilies, now, for the Cuban women, or dolls for the children at the Station
    in Peru—
writing overdue letters,
and now trying to recall every event of the day
for the journal I have promised to keep.

The Journal: Reflections on a Fortnight of Work, March 1900

I condense the story of a fortnight of work
yet it is still so much it overwhelms:

revised, with Miss Cannon, her Classification of the Bright Southern Stars
compared and checked the remarks with the printer's copy
examined, with Professor Bailey, constants for computing observations from
    South America
discussed with the Director the preparation of the Southern Draper Catalogue
measured Algol variables
wrote scientific correspondence
tended to business in Boston
sent out copies of "Standards of Faint Stellar Magnitudes" No. 2
discussed with Mr. Waite his studies and prospects at the Patent Office
interviewed Miss Leland regarding "Faint Stars for Standards of Stellar
    Magnitude"
interviewed Miss Mabel Stevens regarding the Durchmusterung Catalogue
discussed the Director's remarks for Vol. XLV
read and criticized remarks on Professor Wendell's tables of variable stars
discussed notes on observations of stars with Miss Cushman, and instructions
    to render it entirely ready for the printer
received from Mr. King four new chart plates (examining them, I found one
    object showing motion, requiring more study)
read, with Miss M. Stevens, on this object, finding it to be the asteroid
    Fortunata
revised again with Miss Cannon
conducted more letter writing

talked with Professor Bailey about future building

*discussed work with Miss Cushman, looking up several details and starting
     some tables*

*again revised with Miss Cannon and discussed some points about which she
     was doubtful*

examined with the Director the constants for Professor Bailey's reductions

*continued the examination of the four new plates, finding the asteroids
     Urania and Fides*

revised more of Vol. XLV and checked remarks therein

examined and assigned a quality to plates received from the laboratory

looked over, with Mr. Winkley, the Paris charts of the Moon

*discussed with the Director and Professor Bailey the form of some tables for
     variable stars*

*determined with Mr. Bailey the final values for the tables of variables in
     Omega Centauri*

described the work to Miss McGill and started her on it

*found Miss H. Stevens suffering from some anxious attack and sent her home
     in a carriage*

discussed more work with Miss Leland

attended the Observatory Magazine Club in the reception room

reviewed more of Vol. XLV (I hope for the last time)

*examined two sets of photographs of suspected variables, confirming variation
     in both*

measured two "out of focus" plates, with Miss Woods recording

*determined the rate of speed for these observations (65 comparisons made
     with a wedge of shade glass and scale readings given by the observer in 11
     minutes, even with two interruptions by computers who could not wait for
     my attention)*

assisted the Director with more letters

proofed Mr. Wendell's volume

miscellaneous odds and ends and the tying up of loose strands

more work with Professor Bailey

more work with Miss Woods

more work with the Director

more and more and more revising with Miss Cannon

. . .

It tires me just to read the work of the last two weeks.
It's a wonder I have not succumbed
to whatever has overtaken Miss Harriet Stevens,
or to the cold which I feel has got me in its clutches.

The routine work which has to be kept progressing
consumes so much of my time, it seems,
that little is left for the particular investigations in which I am specifically
        interested.

The Director, however, says that my time employed in the above work is of
        more value to the Observatory.
The work is important, and I feel I must continue to show that a woman
        can do it,
all of it,
though many times I must put aside the search for new stars and variables,
the classifying of spectra and the studying of their peculiarities and
        changes.

If one could only go on and on with original work,
life would be a most beautiful dream;
but you come down to its realities
when you have to put all that is most interesting to you aside,
in order to use most of your available time
preparing the work of others for publication.

However, "whatsoever thy puttest thy hand to, do it well."
I am more than contented
to have such excellent opportunities for work
in so many directions
and proud to be considered,
to such a thoroughly capable scientific man as our Director,
of any assistance.

The Journal: A Tour, March 1900

At breakfast this morning Edward announced
that his Chemistry professor and others in the Chemistry department
at the Institute are coming out in the evening to see the Observatory.
I am gratified to know that my son finds my work interesting enough
to share with his friends and professors.
I should just like a wee bit more notice of his plans
so that I might show it in a better light.

My day, however, proceeded as usual, *and with the same amount of work.*
There is never an end to the work on hand
in the photographic and photometric departments.
I walked home to a quick supper and then back to the Observatory,
settling in to odds and ends while awaiting the arrival of Edward and his
company.

They came at half past eight,
and I showed them some of the photographs.
The Director came over from the residence, taking them all through the
building,
telling them about the photographic work and my connection with it.
I then took them to Mr. King,
who kindly volunteered to show them the photographic instruments.
Soon I took the ladies of the party up to the six-inch telescope,
and Miss Cannon showed them the Pleiades and several clusters of stars.

A carriage took the majority home,

while Edward and his friend Neyle walked with me.
Now I have arrived home,
cold from the March evening air, exhausted
but happy for my pride in my son
and for his pride in me.

The Journal: A Personal Reflection on the Value of a Woman's Work, 12 March 1900

During this morning's work on correspondence etc.
I had some conversation with the Director regarding women's salaries.

He seems to think
that no work is too much or too hard for me,
no matter what the responsibility or how long the hours.
But let me raise the question of salary
and I am immediately told that I receive an excellent salary as women's
salaries stand.

If he would only take some step
to find out how much he is mistaken in regard to this
he would learn a few facts that would open his eyes and set him thinking.

Sometimes I feel tempted to give up
and let him try some one else, or some of the men to do my work,
in order to have him find out
what he is getting for $1500 a year from me,
compared with $2500 from some of the other assistants.

Does he ever think
that I have a house to keep
and a family to take care of
as well as the men?

But I suppose a woman has no claim to such comforts.
And this is considered an enlightened age!

I cannot make my salary meet my present expenses
with Edward in the Institute, and still another year there ahead of him.

The Director expects me to work from 9 A.M. to 6 P.M.,
although my time called for is 7 hours a day,
and I feel almost on the verge of breaking down.

There is a great pressure of work certainly,
but why throw so much of it on me,
and pay me in such small proportion to the others,
who come and go, and take things easy?

The Journal: In Which the Activities at Work and Home Grind to an Unpleasant Halt, March 1900

March 14 to March 20—

My daily duties at the Observatory exhausted all my strength each day and at night I was unable to make any record of the days' work and events.

March 21,

Edward unable to go to Boston and in need of a doctor.

On reaching the Observatory I telephoned to Dr. Bailey to call and see Edward.

Tried to work as usual but at 3h 30m had to give up and send for a carriage to take me home.

Also sent for Dr. Bailey.

He came in the evening and told me we had the Grippe.

March 22 to 31

By the time Dr. Bailey arrived, we had a third patient for him—Mary Hegerty had the same sickness.

Only Miss Cannon remained well, so we were left without anyone to do anything for us.

Dr. Bailey found a woman who came in the morning, made beef tea, etc. and looked after us for ten days.

On March 24 Edward was able to go the Inst. for an hour.

On March 25 I was still unable to sit up, but Mary was able to get up and be taken home.

On March 25, 26, and 27 I was able to ride to the Observatory and attend to the work in any department in a general way.
I found the Director confined to the house and also suffering from an attack of grippe.

On March 28 I was able to walk over to the Observatory and in the afternoon I went to Boston.

March 29 and 30 were spent in the usual way at the Observatory.

The End of the Journal, and Very Nearly of Me, 31 March 1900

Sunday at home and with some leisure time to review the events of the past
   few weeks.
All through the winter I have been fighting colds and overworking,
and I do not wonder that at last I had to give up.
During the last half of the month
when I was unable to write anything in the evening
I thought it was due to laziness
which with me would be something heretofore unknown.
Now I can see how hard a fight
the grippe and I had before it overcame me.

I am afraid that in the clutches of exhaustion and influenza
I am less than patient and more prone toward complaint than is my usual
   nature.

I am not sure that this journal
will be of much use to Harvard, or posterity.
But my days, and my thoughts,
were accurately observed.
And as I am engaged in science,
I will obey its tenets.
I will not change the result
though it does not match expectations.

Williamina P. Fleming
273 Upland Road, Cambridge

A Last Reflection on the Journal Accompanied by an Apology of Sorts, 18 April 1900

I find
that on March 12
I have written at considerable length regarding my salary.
I do not intend this to reflect on the Director's judgment,
but feel that it is due to his lack of knowledge
regarding the salaries received by women in responsible positions.
I am told
that my services are very valuable to the Observatory
but when I compare the compensation with that received by women elsewhere,
I feel
that my work cannot be of much account.

Williamina P. Fleming
April 18, 1900

A Nomination, 1900

While I was at home and abed,
Miss Bruce, the generous heiress, was also taken ill.
I am sorry to report that she has passed away.

She will live on in "Brucia," the asteroid named in her honor—
the first discovered by astrophotography;
the telescope named for her—
now high atop a mountain in Peru;
her beneficence to many in the field of Astronomy;
and in the Catherine Wolfe Bruce Gold Medal,
for which, the Director says,
he has nominated me.

A "Field Trip" to Georgia, May 1900

Mrs. Draper insists.
She knows I have been ill,
worries that I work overly much,
tells me I must take time away.

And here is the perfect opportunity:
a total solar eclipse,
which every astronomer knows will occur at the end of this May.
To view the totality,
the entire Sun obscured briefly by the Moon,
we must travel to one of the locations
within the arc of the Moon's shadow upon the Earth.
A small city in Georgia is within that arc.

Mrs. Draper is resolute.
During the last solar eclipse, in 1878,
she assisted her husband with his attempts to observe and photograph it.
But she had remained in a tent, her eyes on a timepiece,
calling out the seconds to totality so that Henry Draper could make his
    observations.
She vows to observe this eclipse first hand.

And the generous Mrs. Draper insists on funding the expedition:
the railways and carriages, the steamer to Savannah, the hotel she knows
    there.
The "crew" of women computers

has been talking of preparations for weeks.
At least half a dozen will make the trip.
Mrs. Draper has invited my son to join us, as well.
(The M.I.T. astronomy professors, I understand,
will have their encampment near our own.)
The Director, vacationing in Europe, had thought to send only his brother,
but has decided to make it a nearly complete vacation of the Observatory
by accompanying us as well.

As I fret over leaving my work, packing my things, closing the house,
I can scarcely recall the newly wed Mrs. Fleming,
packing my trunk to take away to America
and to a life of adventure, prosperity, contentment.

How differently events have transpired!
I have little strayed from Massachusetts
since setting foot here more than twenty years ago.
I cannae claim prosperity,
though when I look at the less fortunate
I find that I am prosperous enough.
But I would claim contentment,
looking around at my snug house and my happy little family.

And now, amidst eminent scientists and their impressive equipment,
away to study the skies,
I will now claim adventure!

An Adventure, May 1900

We board the steamer "Minia,"
and my friend Mr. Adams, Ship's Officer,
shows us the equipment the cable ship uses
to lay telegraph cable across the ocean.
The enormous tank, empty now,
but usually storing miles of coated wire,
dwarfs the man inside it tending to its maintenance;
the brightly lit "testing room," to the delight of our Director,
is a marvel of electricity and communication.

We dine with the Captain and Officer Adams
as well as the ship's dog-mascot,
a sturdy black pup with a white blaze on his chest.
He is happy to be scratched behind the ears by Edward,
and is as well-traveled and learned in "faults, ohms, and grapnels"
as any land-faring man, according to our hosts.

The Officers tell us they will make the crossing again this summer,
when Mr. Adams hopes to see the Paris Exposition,
knowing that I have prepared hundreds of photographs and sent them
    there
for inclusion in a display of modern astrophotography.

We disembark in the port of Savannah
for a stay in a hotel on a lovely square.
Mrs. Draper escorts us on a walk

beneath the enormous oaks draped with curious moss
and we remark on the warm and heavy air
though it still is only spring.

The next day we have a picnic
and games of "India" in the train car
as we travel north to our destination,
the path of totality.

In the Field, May 1900

There being few carriages about, the Director engages a "caravan" of farm
    wagons
to take us out to the vast field where his brother, Professor William
    Pickering,
has been at work for more than a week already,
erecting a set of hasty sheds housing 4- and 5-inch-lens telescopes,
photographic equipment, a camp desk scattered with
charts and journals, ink and paper, maps and drawings and weather reports.

Miss Leland and Mrs. Draper have stayed at the hotel,
but I stand with Misses Wells, Cushman, Eddy, and Woods, the two Misses
    Gill, and Miss Stevens.
(On our "field trip" away, we have been using our given names:
Evelyn and Anna, Louisa and Florence and Imogen and Ida, Edith, and the
    two Mabels,
though the Director remains as formal as ever.)
A reporter approaches, notebook and pencil in hand, and as the brothers
    Pickering answer his questions
(stopping him, twice, from seating himself atop a crate of photographic
    glass plates),
we gaze across the field, a wide horizon well-chosen
to allow us to follow the Sun's progress as it will rise from the
    east-southeast,
and be approached by the Moon until it is overtaken and obscured.

My Edward joins us, having visited the Massachusetts Institute's camp just
    dozens of yards from our own.
He rubs the toes of his boot across the dirt,
bends to gather the rust-colored powder between his fingers,
opens a pocket-knife to scrape at the hard clay beneath,
stands and notices me watching him.
"Och, that'll be the iron," he says, teasing me with the Scottish burr
which I retain but he has nearly lost.
He blows away the dust and wipes his hands with his handkerchief.

Surrounding the telescope sheds, and our party,
are hundreds of rows of knee-high plants,
dark green leaves on sturdy stems—cotton, they say,
though none of us has seen it growing before.
Far across the field
a small crew of dusty workers raise and lower their hoes,
chopping away at the vines growing in red clay along the rows.
They are both men and women,
dark-skinned and light-,
men in loose pants, women in sack dresses, crumpled hats of straw against
    the sun.

I am reminded, here,
with the working women of the Observatory around me,
that women have always worked,
at home or in the field,
forced by poverty or circumstance.
But there are few of us so privileged
as to work among the stars.

Early Morning, Washington, Georgia, 28 May 1900

The Sun has not yet risen, but the hotel is stirring,
window curtains parting, anxious faces peering out at the coming dawn,
hoping for clear skies.

Our party assembles itself in several farm wagons,
some with blankets for the ground, some with hand fans for the heat,
some with shaded glasses for the sun.
We pass a white clapboard church,
farm workers already walking along the road in the cool, still dawn of the
    day,
a well-dressed man driving a mule,
several private carriages passing us on our "hay ride."

As the Sun emerges from the southeastern horizon
revealing only a few thin clouds,
we unfurl blankets onto the patch of ground cleared of cotton plants.
Miss Leland and Miss Cushman volunteer to attend to the Director and
    Professor Pickering,
taking their notes as they pace and plan.

As the Sun climbs
("Do you not think," Mrs. Draper asks, "that we should rather say,
'as our Earth turns her face more fully toward the Sun'?")
the temperature rises,
the fans flutter in my friends' hands,
parasols are opened, tilted toward the east.

There remain more than two hours before the event.

A few carriages arrive, and significant gentlemen and ladies of the city
step out in their shirtsleeves or light-colored dresses
to stroll amongst the amateur astronomers setting up their portable
    telescopes,
to sit upon their folding chairs, and to look about with their opera glasses.

Two women from a literary society carry baskets of soft biscuits filled with
    pieces of salted ham or spread with jam,
passing them out to the members of the visiting astronomical parties.
They invite our group to a county "barbecue" before we depart tomorrow.

A man stands atop a wagon, a Bible in his hand,
preaching to a small crowd gathered beneath him on the red dirt road.

There remains an hour before the event,
and now there is not a cloud in the sky.

Totality, 28 May 1900

A strong voice from the Institute's encampment is the first to call out the
   time:
7 hours, 30 minutes, 0 seconds.
At any second the Moon will appear to reach the rim of the Sun.
The voice continues, announcing every thirty seconds.
Then every ten seconds.

"First contact!" calls Professor Pickering from his telescope,
and a murmur arises from the assembled throng.
Eyes look toward the Sun
through dark sun-glasses or large shards of brown bottles.
Braver, or perhaps more foolhardy souls take furtive, undimmed glances.

It is minutes before the obscuring is readily evident.
It is as if a curved blade is gradually carving away at the upper right edge of
   the disk.
The Moon continues to intrude upon the Sun.
The voice continues calling out the hour, the minutes, the seconds.
But only the astronomers are able to attend to the time,
the note-taking, the camera-loading, the telescope-adjusting.
The people, the rest of them, including, I confess, myself,
seated on our blankets, kneeling beside the rows of cotton, standing near
   the wagons and carriages,
are overcome by the spectacle.

The horizon, 360 degrees around us,

takes on the ethereal deep blue of twilight.

Slowly, color begins to disappear.
The green cotton plants, the blue sky, the colorful quilts on which we sit,
all become drained of color, like Daguerreotype portraits in shades of grey.

There is a sudden coolness,
though the air temperature could only have fallen a few degrees.

Roosters at a nearby farm begin to crow as if they are greeting a second
    sunrise.
Goats in the same barnyard bleat and kneel.
Somewhere in the distance, cattle low.
A church bell starts to peal.

Stars—stars!
And Mercury, and Venus, at mid-morning,
all visible to the naked eye.

And then the call from several scientists at once: "Totality!"

It is as if the Moon is a stopper, sealing the Sun behind it,
while the corona, the Sun's diaphanous flames, seep out around it.
It is, truly, a wondrous sight.

Shadows, 28 May 1900

Totality is just over a minute, in my estimation,
though it is not until the Moon begins to move away
that I once again hear the time called by the Institute astronomer.

"Watch for the shadow bands," calls Professor Pickering,
and it does seem for a moment that shadows race across the ground,
passing rapidly over the white canvas that he has tacked to the side of the
    telescope shed.

The preacher, his Bible closed, sits quietly on his wagon's seat.
The farm workers at the edge of the field kneel, resting in the Moon's shade.

Slowly, the cotton, the sky, the quilts regain their color.

The roosters resume their crowing
the goats take to their hooves,
the cattle in the distance quiet.

"Look between the leaves!"
comes a call from under the lone tree, close to the road,
where some townspeople have set up their picnic.
There, among the shadows cast by the leaves,
are hundreds of crescent shadows—
each space between the leaves a lens
casting its own solar eclipse upon the ground.

The scientists continue their readings,
their observations and measurements,
the hours, minutes, seconds still to be recorded,
their voices just as urgent.

As the Sun regains its shape,
the crescent shadows disappear,
the blue twilight of the horizon lightens to the day.

I look toward Mrs. Draper, who sighs contentedly.
"Well, Mrs. Fleming," the benefactress asks. "What did you think of that?"

"Och, Anna!" I say. It is difficult to find my voice after such a spectacle.
"I am so glad to have seen that!"

Final Contact, 29 May 1900

We are invited, next day,
to the grounds of a country church
where a large cast iron pot is stirred
and whole pigs are roasted over low smoking pits.

There is not a church service,
though our hosts tell us that some of the congregation
spent the eclipse on their knees, praying for redemption,
afraid that the minutes' darkness would portend the end of the world.

Our hosts are a literary society,
civic leaders proud of Wilkes County, Georgia's role in such a scientific
    endeavour.
The women who serve us—wives of the civic leaders—
express surprise that we Harvard women are not, in fact,
wives of the Harvard scientists, but scientists of a kind ourselves,
furthering the knowledge of the heavens,
though in truth we worked less during the event yesterday
than we would have in the Computing Room in Cambridge.

After our meal (the pork and a stew of chicken, bacon, corn, beans, and
    tomatoes, and more of the biscuits we enjoyed yesterday),
on our way to the train depot, our carriages take us back along the cotton
    field
where Professor William Pickering will stay briefly behind,
directing the crew who carefully pack the telescope parts,

disassemble the sheds, gather the notes and instruments
and exposed glass photographic plates, carefully wrapped and kept from
    the light.

The instrument readings were telegraphed immediately following the
    event—
the precise times of contact and totality, the Sun's and Moon's locations in
    the sky,
the readings of the thermometer, anemometer, magnetometer—
all sent along telegraphic cables from the fields along the path of totality to
    New York City,
then along the depths of the Atlantic Ocean,
received by astronomers in London just sixty seconds later,
and from there to nearly every important observatory in the world.

We, however,
do not travel as fast as electrical pulses along a wire.
Sitting on benches in the shade of the depot,
we adventurers await the train to begin our trip home.

Back to Work, June 1900

The telescopes have not been idle.
The assistants who remained in Cambridge for the solar eclipse
have continued their systematic photographing of the stars;
Mr. King has developed a supply of photographic plates which patiently
    await our scrutiny.

There are photographs from many points along the path of totality,
each major observatory having set up its own viewing "party."
And a good thing that is.
Professor William Pickering's solar eclipse photographs, unfortunately,
could not be developed satisfactorily.

His equipment, it seems, was jostled at the least opportune moment.

Aside from the Sun and the stars,
there is Eros to prepare for—
the tiny "planet" whose path I managed to find on our historical plates,
and which should be passing through our skies before the year is out.
Along with the readings from the eclipse,
in Georgia and in other locations along its path,
little Eros may help to finally determine
just how distant our Sun is from our own planet Earth.

And, the Director reminds me, we have such a backlog of data
gleaned from our photographic plates

that he estimates it would fill twenty-eight new volumes of our *Harvard
    Annals.*
"You must put it in shape for publication," he says.

I do not imagine that he often considers
how many tables, how many columns of figures,
how many explanatory remarks,
how much time and effort, that entails.

"This must be your priority, Mrs. Fleming," he says.
Back to work, indeed.

Lady Assistants, Autumn 1900

I perhaps do not need to mention
that I am <u>not</u> awarded the Bruce Medal for 1900,
despite the Director's nomination.

The Director, however, has been recognized for a different award,
his second gold medal from the Royal Astronomical Society,
for his advancement of astrophotography and his work with variable stars.

I am the discoverer of many of those stars.
And Mrs. Draper is the founder of the research.

What, I wonder, would we be required to do,
Mrs. Draper, Miss Maury, Miss Leavitt, Miss Cannon, myself,
to be considered "astronomers"?
We are benefactor, computer, assistant, curator,
often qualified with "lady" preceding our title.
What privilege and honor might come our way,
what opportunities might we be afforded,
if we, reflecting our constant study of the stars,
were known instead as "astronomers"?

Well.
We are kindly sent a copy of the speech given in London
announcing the Director's well-deserved prize.
Mrs. Draper is credited for making the work possible.
I am noted as *"that most careful observer"*
of the Director's *"lady assistants,"* who *"has been able to announce*

some of the most interesting discoveries . . . that have ever been recorded."

I am beginning to find that "lady assistant"
does not quite encompass my work
or my accomplishments.

Miss Cannon's Classification, 1901

Finally, Miss Cannon's catalogue of Southern stars is ready for the printer.
Not since Miss Maury
has my time been so dominated by another person's work:
assembling the many-columned tables,
prompting the author for her explanatory remarks,
editing and reading the proofs.

Yet it is Annie Jump Cannon's own work:
a system built upon my classification of stars by their spectral absorption
    lines,
and also upon Antonia Maury's complex classification of inter-
    relationships among the stars.
She has taken my labels of A through Q,
eliminated those rarely used since its inception,
and re-ordered the most useful ones.
They reflect, she thinks, the stages of a star's development,
though this is still in question.
She places first the strongest spectral lines, designated O,
and follows with a new order:
B A F G K M.

Perhaps a star is born bright,
and dims and cools with age.
Or perhaps it begins an ember,
and grows to a fiery ball as it matures.
We do not know which, if either, is the answer.

Our attempts at ordering the stars—
Stecchi's, mine, Miss Maury's, Miss Cannon's—
are merely the vocabulary we use
as we work to understand them.

A Graduate, June 1901

He is the first in his family—the only, to now—
of the Stevens, of the Flemings
(as much as I knew of James, at least)
to achieve a college degree.
And now my son Edward Charles Pickering Fleming
is a graduate, an engineer.

He will join a mining company,
studying the metals and minerals of the Earth.
He will be going West.
"Just a few days' train ride away, Mother," he says.
(It seems I am "Mother" now, "Mam" lost to his childhood.)
"Or, thanks to Mr. Bell, perhaps a telephone call."

Now I must again become accustomed to being without him.

My brother Charles, in celebration,
raises a glass to those assembled for the occasion—
his wife and sons, our mother, our brothers and sister and their families as
    well—
and says, "Look to the horizon, there.
Above it, there, the sky, belongs to our Mina.
And below it, here, the Earth beneath our feet,
that now belongs to my handsome nephew Edward.
Slàinte!"

I raise my own glass to my son,
my eyes shining with pride.
"Slàinte!"

Another Nomination, 1901

"Well, Mrs. Fleming," the Director says,
"'if at first you don't succeed, try, try again.'"
He has put forth my name, again, for the annual Bruce Medal.

"I argue again that you have discovered more variable stars
than any other person, man or woman," he says,
"though I make note that you have had fewer discoveries this year
due to your time spent preparing the *Annals*."

I perhaps do not need to mention, again,
that despite his efforts I do not receive this award.
But I do thank the Director for the nomination.

The Grippe, 1901

I recognize the fever,
the cough, the ache, the weakness, the fatigue, the labored breathing.

I tell Miss Hegerty and Miss Cannon to stay away,
and send word to the Observatory.
I salt a soup bone and an onion in a pot on the stove
and keep the broth hot to sip for strength.
Then I wrap myself in a quilt,
some camphor in a flannel around my neck,
and take to bed.

It is nearly three weeks before I am recovered.

When I return to the Observatory,
I find that the Director and Mrs. Pickering and nearly half of the women
    computers
have been similarly infirm.

We can look into the universe
and see heavenly bodies so many miles away.

What a shame it is
that we cannot look into our own bodies well enough
to relieve the suffering contained there.

An Anniversary, 1 February 1902

Mrs. Draper, Mrs. Pickering, and I have been planning for weeks:
a fete for the Director on his twenty-fifth anniversary in the position.

We call him to the photographic library
to look at "some particularly interesting plates" and surprise him.
Lizzie Pickering provides a tall silver loving cup,
and Mrs. Draper presents him with an elegant mantel clock.
We computers have taken up a collection for a new desk chair.

There is a celebratory luncheon with amusing speeches
and the Director, usually opposed to "events," seems genuinely pleased.

Och, I have complained about the tasks assigned to me,
the compensation unequal to the responsibility.
but I am ever aware that this Director has shown other men
the work that women can do.

Had I begun to work, twenty-three years ago,
for another man,
I might still be working as a maid.

Dinner with Astronomers, 1902

I am invited to become a member
of the American Astronomical and Astrophysical Society,
of which our Director is already a founding member and officer.
I am delighted to accept.
There are few professional organizations which open their full
    membership to women.
I am grateful for the opportunity.

There will be a conference in Washington—
speeches given and papers read, a dinner.
I am more than happy to attend the lectures,
discuss variable stars with my colleagues,
extol the virtues of astrophotography.

But what of the dinner?
Although I am not the only woman member,
I must wonder whether the men of the society, the vast majority,
really welcome women to their social events.
Will any of the other women assert their right to attend?
Would we be hurting our cause, our hopes for full inclusion,
by pressing this small advantage?
And,
would I need to have a new dress?

Today, from the AA&AS President Newcomb,
I receive a reply to my cautious R. S. V. P.:

"Dear Mrs. Fleming,
I am much disappointed to notice
that although you hope to be here at our meeting,
you do not propose to join the dinner.
Possibly you may be under a misapprehension,
supposing that the dinner is only for the men of the society.
Permit me, therefore, to assure you
that all members are equal,
and that we should like very much
to have our lady members with us."

How refreshing! How reassuring!
And now I must decide about a dress.

Building, 1902

With funding for capital improvements, a new wing is being added to the
    Brick Building,
our repository of glass photographic plates,
the collection of which I am Curator.

It is a modest addition,
of plain red brick, thirty feet square and three stories tall,
enough room, at our current rate,
for at least ten more years' photography.

Importantly, a fire hydrant has been installed.
My greatest fear,
the destruction of the glass plates by flame,
is somewhat lessened.

Now to the next ten years of work!

A New Benefactor, 1903

Mr. Andrew Carnegie—
titan of industry, builder of libraries, champion of immigrants—
is now also
benefactor of the Harvard College Observatory.

His grant will allow us to hire new computers.
"Ten, Mrs. Fleming. Find us ten more women
capable of recording data, computing the mathematical reductions,
investigating the plates," the Director instructs.

I turn to the file of applications,
and consider the women who have written to me,
spoken to me,
pleaded with me.
If only I had a place for each of them!

A Provisional Catalogue, 1903

Miss Annie Jump Cannon's latest achievement,
a catalogue of variable stars, lists 1,227 of them,
half of those discovered at Harvard.
I am listed as the discoverer of 166.

Before I began here,
before astrophotography so expanded our view,
fewer than 150 variable stars had ever been identified.
I wonder if Mr. Daguerre had any idea
that the tools of a family portrait
would someday so expand the universe.

Och, I wonder if my father,
in his Daguerreotype studio in Dundee—
sitting me in a chair, lifting my chin,
reminding me to be still while he exposed the photographic plate—
had any idea that the same tools
would someday so expand mine.

The Return of Miss Leavitt, 1903

Like Miss Maury, Miss Henrietta Leavitt cannot be contained.
After several trips to Europe,
time with her family in Wisconsin,
college teaching,
she has returned to the Observatory,
focusing her attention on the Orion nebula—
a foggy nest possibly filled with variable stars.

I wonder whether these brilliant women
(Miss Cannon included, though she seems a bit more grounded)
would be more content if they,
like the few male astronomers of equal intellect,
had the authority of their own laboratories, observatories, staffs of
    assistants.

Or would such responsibility, once bestowed,
only restrain their imaginations?

Invention, 1903

Miss Leavitt has made an invention:
she has taken some of our glass photographic plates
of the guide stars which we use to measure magnitude,
and has had the glass cut in smaller pieces,
framed in metal,
and fixed to long handles.

The resulting tool looks something like a fly-killer,
of the type used on front porches and at summer picnics.
But she cleverly calls it a "fly spanker,"
too small to do much damage to an insect.

She distributes the "fly spankers" about the Computing Room.
Their actual usefulness is immediately apparent.
As we view our photographic plates on their viewing frames,
we need only hold the tools next to a particular star,
choose the guide star most similar in brightness,
and measure magnitude from that.

It is remarkably faster,
and possibly will prove more reliable, as well.

With her tool,
Miss Leavitt has quickly confirmed sixteen Wolf-type variable stars
and found fifty new ones.
I have verified her discoveries myself.

Science advances,
and flies suffer no harm!

An American Citizen, 1904

I so enjoyed having Edward home
for the Christmas holiday.
He is a mining engineer,
working at the Copper Queen property in Arizona.
He insists I visit him there, some day—
the landscape there so different from anything here
or in Scotland
that he says I must see it.

I made sure to have my mother here,
so he might see her.
At her age, now,
one must be careful in how one takes one's leave.

Now the house is quiet again
after evenings of hot cocoa and cards,
visits from his school friends, groupings of family.

He has applied, he says,
to become a naturalized citizen of the United States.
It will make it easier for him to travel for his employer,
as he is likely to do in the future,
wherever the metals needed for industrialization may be found.

And now that he has seen so much of the country,
he feels he is a part of it.

He feels, I think,
a belonging to it.

He is healthy, a bit tanned, enthusiastic.
And now he is an American.

Room for More, November 1904

Sudden and unexpected,
and all the more tragic for what she leaves behind,
my brother Charles's wife Eliza
has suffered a stroke and died.

Two boys, eight, and twelve, without their mother.
Charles, of course, bereft.

Their family, as so many,
has endured misfortune already.
Two daughters, little Gladys Williamina never reaching two,
and another, wee Mina Fleming Stevens, living only weeks.

Their house too full of sadness,
and too difficult to manage without Eliza,
Charles and young Charlie and Malcolm will live with me.

There is enough room in my house.
Though my mother is here now,
Miss Cannon has graciously left to stay with her sister,
and my Edward is off on his own.
There is room for my brother and nephews.
And there must be room for sadness, for a time.

I must do what I can
to make sure there will be room for happiness again.

Losses, 1905

Late tonight,
the snow falling outside quieting the household,
his boys and our mother abed,
Charles and I sit by the stove, pour the tea, discuss the loss.

His lads are resilient—
missing their mam, but so busy with being boys
that their loss is one felt in swaths, not constantly.

Charles misses his wife like a limb,
my home the crutch he leans upon.

Did I ever miss James? he asks.
For a short time, I say.
Desperation and resentment helped to forget the life I'd hoped we would
    have.
When I finally learned of his death—
our sister Mary in Dundee (herself a widow) had heard of it—
I was sorry only for his mother.
Not for me, who had learned to be happy without him.
Not for Edward, who never knew him.

We mourn another loss, two years ago:
our brother Robert's son Albert,
from the slow and saddening certainty of consumption,
until the tuberculosis finally took him.

He was only the same age as my son.

Our father, I tell Charles,
is a loss I still feel,
though Charles was too young to remember much of him.
I see him, though, in the faces of my brothers,
even our sons, and in the work they do—their craftsmanship.
"And in you," Charles says, "in your curiosity, I suspect."

More tea, and the conversation turns quietly to our mother,
still with us but slipping gradually away.
She suffered her own losses:
our father, our two brothers who did not survive childhood,
her mother our grannie.
These days she requires assistance herself.

I had never told Charles of the miscarriage I suffered before Edward,
but I do so now that we are sharing our sorrow.
That was a loss that made my abandonment more difficult to bear,
knowing that James left me to face pregnancy alone.

We are both past weeping
and tomorrow we must get his boys off to school,
our mother prepared for her day here
(a nurse comes to tend her for a time),
and ourselves to work—me to the Observatory, Charles to his workshop.
He banks the fire, I rinse the teacups,
and we bid each other a good night.

Nebula, 13 January 1905

Today, on one of the Northern plates, I find something new:
a wispy triangle,
a tangle of filaments
in the Veil Nebula of Cygnus.

If only I could travel through the "ether,"
this beautiful wisp
is something I would surely like to see.

Once Again: Cambridge, 1905

So.
Here we are.
They say my name has been put forward again
for Miss Bruce's Medal,
the coveted prize in Astronomy.

Still, I am an unlikely choice,
being a woman,
though Miss Catherine Bruce herself suggested
that women should be eligible for the prize.

The Director has shown me the form
on which he has written my name in nomination:
"Mrs. W. P. Fleming,
for her discoveries and continuing researches in stellar spectroscopy,
extending over the last 24 years.
She has discovered nearly all the Novae,
stars of the 5th light,
and stars having hydrogen lines bright
which have been found during the last 20 years."

It seems that even the Director and his diplomacy cannot overcome the
    learned men,
with their mansions and their clubs and their cigars,
who cannot imagine that a woman,

who does the work that they themselves do,
as well as the work that their wives do,
deserves such a prize.

And so I do not receive the Bruce Medal.
Och, of course I did not expect it.

I certainly do not require it to continue my work.
But it is gratifying to know that the Director, at least,
believes that I deserve it.

At Wellesley, 1905

They are young women in a setting
in which I wish I'd have been,
when I was a young woman thirty years ago—
in a respected women's College,
in a Physics classroom,
with a capable female Professor.

Professor Sarah Whiting has invited me to speak to her class, again,
on new variable star discoveries
and astrophotography and spectroscopy in general.
Professor Whiting teaches through experimentation and inquiry,
in the style of our Director, whose classes she observed.
Her discussions are always lively—
her students are curious, outspoken, full of thought-provoking questions.

Today I am rather taken aback by a question
from a young woman with a pencil in her loosely coiffed hair.
"You say, Mrs. Fleming, in your address 'A Field for Woman's Work in
    Astronomy,'
that we women should 'labor honestly, conscientiously, and steadily.'
It seems we should expect to do whatever labor is handed to us,
rather than direct our own investigations.
Do you mean that this is all we can aspire to,
as women in the sciences?"

"Why, no," I answer.

I pause, trying to collect my thoughts.
"I suppose, when I wrote that a dozen years ago,
I meant that astrophotography allowed me, and the women under my
    charge,
to study photographic plates and to form conclusions from them,
independently of the men whom we assisted.
What I meant, I suppose,
is that when there is photographic evidence,
the stars speak for themselves.
We have only to observe them."
I continue, "Since then, though, I would say,
more women have entered the field,
and the very choices of what to observe—
where to point the telescope, as it were—
and how to conjecture on what the stars tell us,
beyond their brightness, size, and distance, is being done by women,
like your own Miss Cannon."
I look toward Professor Whiting, who instructed Annie Cannon when she
    was a student here.
"She 'mans' the telescope on her own, work once proscribed to women,
and develops her theories through her own research,
not as an assistant providing that research to a man."

I turn back to my questioner,
and see in her the impatience
with the limits on our sex
that I have often felt I must accept.
"But there is more work to do," I say,
"in astronomy and in all work for women.
I sense that you are apt to do that."

My questioner nods and sits,
satisfied, it seems, with my answer,
but perhaps not quite satisfied with the challenges ahead.

Taking a carriage home, I consider the challenges ahead of me.

If only I could find the time
for more of my own investigations,
to study the plates and compare them to the historical record,
to develop my own theories and explore them,
to be not only the assistant to the Director,
the supervisor of women computers,
the Curator of Astronomical Photographs.
If only I could be an Astronomer.

Corresponding with Mrs. Carnegie, 1905

"Perhaps a letter from you, Mrs. Fleming, would be of benefit," the
    Director says.
He has spoken to Harvard's President Eliot about Mr. Andrew Carnegie,
the steel magnate whose largesse has not been renewed.
(I have had to let go some of the newly-trained computers,
though I hope most will be back if their funding is restored.)
He has invited Carnegie to the Observatory, hoping to impress upon him
how well his previous contribution has been put to use.

"To Mrs. Carnegie, one woman to another," the Director continues.
"They have a daughter, I believe.
Explain the important role of women here.
Perhaps," he says, raising an eyebrow,
"you can might add a bit of that Scottish burr we're so fond of to your
    writing.
Remind her that you are a countryman of her husband's.
'Countrywoman,' I should say."

Och, imagine me corresponding with the family of the great man,
the self-made millionaire, the immigrant philanthropist!
I find the addresses in our files:
the mansion in Manhattan, the holiday cottage in Florida,
the castle in the Scottish Highlands.
I carefully practice a draft,
then use my finest ink and paper
and begin my correspondence with Mrs. Louise Carnegie.

Star Library, 1905

To Mrs. Carnegie I describe the Astronomical Plate Collection
as a library, and myself as its librarian.
Her husband is much admired for the many community libraries he has
  fostered,
the repositories of knowledge and story
made available to hard-working citizens.
Mr. Carnegie, himself, was once a poor young lad allowed access to a
  library.
He credits this for his success.

In his correspondence with Mr. Carnegie,
the Director has likened our collection to the Sibylline Books,
the collection of oracles kept safe and secret
for consultation by the kings of ancient Rome.

I have not aimed quite so high,
but I have described how an astronomer,
anywhere in the world,
might consult with me about our collection;
how I can find for any scholar
those stars that appeared
in a particular location on a particular date;
multiple dates and photographs of a particular star or comet or nebula;
and how I have identified many of those myself.

I ask after her husband, her daughter,

and wonder when they were last at their estate in The Highlands of
    Scotland.
I tell her that I miss my homeland
and how, should there ever be the time,
I should like to see the country again.

I show the letter to my brother Charles,
hoping he'll alert me to any errors.
He only reads and whistles.
"Och, you've come up in the world, Mina," he says.
"Imagine—Carnegie, the man himself!"

A Reply, 1906

"We rejoice," Mrs. Louise Carnegie writes,
"that a daughter of bonnie Scotland
is doing such grand work!"

The gracious Mrs. Carnegie and her husband
met with University President Eliot in New York,
but did not have time to travel to Cambridge.
Perhaps, she writes, without any particular date,
I might lunch with her in Manhattan one day,
even visit them at Skibo Castle, should I return to Scotland!

Their daughter Margaret, poor thing,
is missing her school,
at home with a badly sprained ankle.
Do I have children? she asks.

I am planning my next letter, answering her questions,
adding something to cheer little Margaret.
I hardly notice
that there is no mention
of renewed funding for the Observatory.

A Sky Map, 1906

Miss Leavitt, with her fly spanker and a new technique
of pairing a negative glass plate with its positive and peering through them
    both,
has found a wealth of stars—hundreds!
in the Magellanic Cloud.
At this rate, Mr. Bailey and the Director believe,
there must be thousands more stars than were ever imagined.
Harvard can assemble a complete Map of the Sky,
verifying new plates with each other and with historical plates and
    catalogues.
He has encouraged Miss Cannon and Miss Leland to copy Miss Leavitt's
    method,
divided the sky into three parts,
and assigned one to each of them.

Such a map will ensure that our "library" will be the envy—
and the authority—of the astronomical world.

College Women, 1906

I am honored to be named "fellow" at Wellesley College and Radcliffe,
    among others,
having taught as a guest so often
in their classes for women in Astronomy and Physics.

I am equally honored to know the women
who have worked so diligently to provide these opportunities.

Professor Sarah Whiting of Wellesley
is an inventive instructor,
tirelessly designing and equipping her laboratory as well as her observatory,
devoting herself entirely to educating women.

Radcliffe President Mrs. Elizabeth Cary Agassiz, on the other hand,
is centered by her family, raising children and grandchildren
and creating places for women to be educated where none had previously
    existed.

Their students,
some of them now assistants under my charge,
including Annie Jump Cannon and Henrietta Leavitt,
are sure to be important contributors to the field.
I am honored to be among these women,
to be their "fellow" scientist.
I am happy also to be their friend.

Valentines, February 1906

I make hot chocolate and we sit round the kitchen table
with scissors and paper and paste,
little Charlie and Malcolm, my brother Charles and I,
crafting clever little cards for the young lads to take to school.

The boys make a card for their Grannie, looking on from her corner,
and Charles proposes one for Miss Honan,
who has come to help our mother when we are all at school or our work.
I slip a folded heart into a letter for Edward and address it for post.

I make a card for dear Margaret Carnegie,
who has sent me a paper flower Valentine.
Her mother and I have exchanged a number of letters,
and I have sent one of my dolls to her daughter.
Margaret is still at home from her school with a slow-healing ankle.
I am sorry for her, knowing the pain.
My own ankle was crushed by a delivery cart in Dundee
when I was just a bit younger
than Margaret and my nephews are now.
It was only my obstinate father, as I remember,
crafting a boot reinforced with steel braces,
that allowed me to keep my foot,
rather than the amputation that the doctor proposed.
He did not live to see me walk again unaided.

The boys and Mother abed,

Charles and I paste errant paper hearts back into place,
leave the Valentines out to dry on the table,
and sip the last chocolate from our cups.
Though we never forget our losses,
there are still evenings, like this one,
that bring happiness.

An Honor, May 1906

"A word, Mrs. Fleming," the Director says,
"when you have a moment."

I have been sitting with Miss Cannon,
discussing the remarks for her next catalogue,
Miss Leland and Miss Leavitt at their light lecterns,
fly spankers in hand, examining their plates,
the other assistants at their reductions.
The women look up, curious.

I speak again to Miss Cannon,
careful to face her and speak clearly,
as her hearing is such a challenge.
"A few more remarks," I say, "and you will be ready for the printer."

I climb the stairs to the Director's office,
where he sits at his many-sided desk.
He looks stern, as he has lately,
but smiles as he indicates a letter.

"I have been informed," he says,
"that the RAS wishes to induct an honorary member.
It is you, Mrs. Fleming."

He notes I will be only the third woman—
and the first woman in America—to be so named

(honorarily, as full membership is restricted to men).
It is a great achievement, he says,
and a feather in the cap of the Observatory, to be sure.
He rises and shakes my hand.
"Congratulations, Mrs. Fleming. It is well deserved."

I cannae conceal my smile as I return to the room,
my colleagues looking up expectantly.
"It is the Royal Astronomical Society," I tell them.
"I'm to be a member. An honorary member."

"Oh, Mina," Miss Leland says,
and the others crowd round, offering congratulations and embraces.

"Och," I say, "It is not so momentous."
But I return the embraces, accept the congratulations, and sigh, smiling.
The Royal Astronomical Society.
Perhaps I have begun to join the ranks of the Astronomers.

Mrs. Pickering, August 1906

The Director does not show his sorrow,
but it is evident in his absence from the Observatory,
the quiet service held in the residence with only the household present.

His dear wife Lizzie Pickering has died.

I remember her kindness to me,
the way she cultivated both the gardens around the Observatory
and the society friends who helped it to grow.

She was always a charming hostess,
but in the last year she had withdrawn.
A badly broken ankle would not properly heal;
recent surgery did not assuage her constant pain.

Though we would express our sympathy,
we computers respect the Director's privacy.
Several of us go, on a Sunday,
by carriage to Mt. Auburn Cemetery
and leave some of the flowers she loved so much on her grave.

We go about our work, continuing to map the heavens
which must surely be more populated now.

A Fellow, 1906

I am asked to visit the office of one of the American members
to fill out a membership card and to be interviewed
for the Royal Astronomical Society.
(The official meetings are generally held in London, of course,
so I will send my acceptance from Cambridge.)

In the anteroom I remove my gloves
and wait for the clerk's attention.
"Yes?" he asks.

"I am to provide you some information," I say.
"I am Williamina Paton Fleming."

"Information?" he inquires.

"For membership, I believe?
I am to be an honorary fellow of the RAS."
I pause, hesitant to continue, but take a deep breath and press on.
"I am an astronomer."

He stands. "Ah! Yes! Mrs. Fleming! My apologies."
He shakes my hand, indicates a seat, fumbles around his desk.

He asks my title and address,
but seems befuddled by a troublesome word.
"Eh," he begins, embarrassed, "perhaps I should just . . ."

I peer across the desk and see him circling his pen
above the word "fellow."

"Change to 'member,' perhaps?" he asks.

"Och," I say, "I think it is hardly so specific
to the male of the species.
One might be a 'fellow traveler,' or a 'fellow citizen,' do you not think?
I believe I might be termed a 'fellow' without offense.
To myself, at least," I add.

"Of course," the clerk sputters.
He takes the remainder of my information, hands me a leaflet.
"Thank you, Madam."

I put on my gloves.
"It was my pleasure," I say.

Publicity, 1906

The Director, following a sad absence, now seems back to himself.
His zeal for the cause of astrophotography has returned.
He strides to my desk, brandishing a folded newspaper.
"We must take advantage, Mrs. Fleming," he says.
"The women of the country—and the men as well—
should know of your work here."

"Och," I tut. "I may enjoy credit for a few discoveries,
in the *Annals* or the star catalogues,
but I certainly do not crave fame."

"But it is for the Observatory.
For all the women under your charge.
For those intelligent young women
who might entertain the idea of pursuing the sciences.
And," he adds as if it is an afterthought,
"publicity of your accomplishments
may also serve to attract donors to our 'stellar' cause."

Ah.

"I have spoken to the *Globe,* Mrs. Fleming.
Please speak with their reporter
if he or she should call on you."

Of course I acquiesce, though I hardly know

when I shall find time for an interview.
So, when she does call,
I agree to meet the reporter at the Public Garden
on an afternoon otherwise devoted to errands in Boston.

I rather impatiently answer her questions
about the errands I have been attending to,
the family from which I came,
the occupation of my little leisure time;
until we come to the variable stars and nebulae,
the honorary fellowships, the lectures at the colleges,
the publications of star catalogues,
the astro-photographic plate collection.

The article, when it
appears in the Sunday edition,
does list my scientific accomplishments
as well as the work of the other women here.
But the Observatory staff, next day,
are amused when they read the rest:

"Doing?
Why, I have been to the tailor for a suit.
I wish it might be a genuine, tailor suit, too.
Here they have made me a 'Fellow of Wellesley'
and a 'Fellow of Radcliffe'
and a 'Fellow' of two or three more colleges
where I am doing a man's work
with a woman's pay.
I'd like sometimes to get a 'Fellow's' clothes,
salary and all, if it were possible!"
and Mrs. Fleming's brilliant brown eyes flashed
and her voice was very earnest,
though her sentence ended with a laugh.

Och. I had only been talking,

in the middle of a busy day,
not intending to deliver such a speech.

The director, passing through, pauses just a moment.
"I see, Mrs. Fleming, that you've granted your interview."

"Aye," I say.
I believe that I detect the hint of a bemused smile on the Director's face.
Perhaps, also, for just a moment,
my "brilliant brown eyes" flash.

Headlines, 1906

The newspaper clippings are collected by the Librarian at Gore Hall,
or sent to me by post from stargazers around the country.

They are from cities both close and distant:
Boston, Massachusetts; Windsor, Ontario; and Arcadia, Florida;
Champaign, Illinois; and Galena Kansas;
Elyria, Ohio; and Kenosha, Wisconsin;
Red Rock, Colorado; and Beatrice, Nebraska.

Their newspapers go by the names
Herald, Reporter, News, Telegraph, Argus,
Opinion, Advocate, Leader, Citizen,
Tribune, Express, Sentinel, Globe, Sun, Star.

The headlines, to my mind,
in their large type and capital letters, are alarming:

A FAMOUS ASTRONOMER.
Mrs. Williamina Paton Fleming is
a Brilliant Scientist.

*SHE DISCOVERS STARS. Mrs.
Williamina Fleming's Fine Record
as an Astronomer.*

*WORLD'S   GREATEST
WOMAN ASTRONOMER. In
the Past Twenty Years Mrs.
Williamina Paton Fleming, of
Harvard College, Has Discovered
Over One Hundred Stars.*

*HONOR   AMERICAN
WOMEN. English Scientific
Society Elects Lady Astronomer to
Membership.*

*CLEVER   WOMEN   WHO
STUDY THE STARS. "They
are remarkably well adapted to
astronomical work by reason of
their clearsightedness, physically as
well as mentally, their natural
carefulness and attention to detail,
their delicacy of touch, all of
these and many other attributes
of woman combine to make her
exceedingly valuable in the work of
the observatory."*

WOMAN DISCOVERED SIX
OF THE NINE NEW STARS.
IS A CURATOR AT HARVARD.
Has Been Made a "Fellow" by Many
Universities. She Likes Housework,
Too.

Och.
I scarcely know what to do with it all,
but to hope that it results in support for the Observatory,
and to get back to the work at hand.

Miss Cannon's Catalogue, 1907

It is quite an undertaking, cataloguing every star—
its magnitude and location,
its type, its spectra,
its discoverer.

Miss Cannon does her work in near silence,
her hearing so damaged by illness
that she must read the lips of her companions in conversation,
or hold to her ear the sound amplifier that she keeps at hand.

I bother her to finish her remarks,
to explain every anomaly, every error, every unexpected result in her tables.
but she is most capable of working on her own.
She remains the only woman manipulating the telescopes here,
turning the gears and pulling the ropes to follow her stars,
working quite alone, often, in the dark of the night.

Her latest *Catalogue of Variable Stars* is finally published;
the first was *A Provisional Catalogue of Variable Stars,*
then one revision, then another.
For this, we have decided "provisional" is no longer descriptive—
the discovery of new stars is so rapid now,
every catalogue is "provisional,"
out-of-date before it is through the printer's,
like a photograph of a child grown days older
before it can be framed and set upon his mother's table.

A Marriage, 1907

"She is good with the boys," Charles says,
"and at keeping house, and nursing, and cooking,
with a good head upon her shoulders,
and kindness,
and a fondness for us, I think."
He looks away, my forty-three-year-old brother, blushing.
"It has been more than two years, Mina.
What do you think of it?"

He is talking of marriage, to Miss Honan—
Josephine, her given name—
who has been caring for our mother here,
and cooking, often,
and watching over young Charles and Malcolm.
They will take their own house, nearby, I hope,
to the benefit of all.
(Except, perhaps, for me.
I will miss the sound and presence of my brother and my nephews,
and Josephine, who has become a friend
as well as a helper to me here.)

"I believe your dear Eliza would want nothing more
than to know her husband and sons
are well cared for, and happy," I say.
"But please, Charles, don't take them all too far away."

"Och," he says, "we'll stay on a bit,
and then find a place 'round the corner.
You'll not be rid of us so easily, Sister."

It is the next day, Charles late at his shop,
while we are putting supper on the table,
that Josephine speaks of it to me.

"I am so fond of them all, Mina," she says.
"And he'll stay with me, do you not think?
Should there be difficulties, I mean, or—
or more children?"

Her question sends a moment's panic through me,
recalling my own abandonment,
what might have become of me, of Edward,
if circumstances, and the Observatory,
had not favored us.

"Of course," I tell her.
I think I know my brother well enough to promise her this.
"Yes. I think you will be very content."
I embrace my new sister-in-law
and think, perhaps, there will soon be more children.
"Josephine," I say, "Sister, I am very happy for you all!"

Fire, 4 March 1907

It is the nightmare that recurs, it seems,
whenever I have read in the newspaper
of a house, a factory, a business, a school, a church
consumed in flame.
I wake in the dark,
imagining the smell of smoke which dissipates
as I sit up in my bed,
wrap a shawl around my shoulders,
sniff the air, and listen, waiting for my head to clear
and I can be sure that it is but a bad dream.

It is not at my house where I most fear a fire, however.
It is at the Observatory, in the wooden cupboards of the plate collection,
the glass plates in their brown paper covers
alight, crackling,
the history of the stars melting away.

So, at three minutes of six today,
when we were deep in the maze
of some manuscript being readied for the printers,
when the telephone alarm rang in sevens—the code for the residence—
we leapt to our feet,
the director picking up an extinguisher,
and we hurried down the stairs, gathering the staff as we went.

At the door to the residence
I grasped a ready extinguisher and, entering the scene,

found it full of smoke,
a man there already spraying water on a lounge sofa,
and flames spreading along the floor
and climbing the wall above a set of windows.
I trained my extinguisher on the casement and the cornices
while the others must have extinguished
whatever was aflame in the parlour below.

In just a few minutes all the flames were out
and we stood a bit unkempt and damp,
smelling of smoke ourselves,
when the horse-drawn steam brigade arrived
and the firemen saw to the rest.

I will write it into an exciting story
in a letter to Miss Margaret Carnegie,
fashioning the Director and myself as unlikely hero and heroine
arriving in just the nick of time to save the history of the stars.
I will also remind the assistants,
who grumble over our monthly alarm practices,
that our quick action
has already saved ourselves and the plate collection
from quite a great loss.

Marvelous Work, 1907

"Many astronomers are deservedly proud
to have discovered one variable
and content to leave the arrangements
for its observation to others.

the discovery of 222,
and the care for their future on this scale,
is an achievement bordering on the marvelous."

I read Mr. Turner's words of praise again.
He (Herbert Hall Turner, of Oxford and the RAS)
is far too generous.

But I am proud of this work,
my *Photographic Study of Variable Stars.*
With the aid of Miss Leavitt's "flyspankers"
I have tracked my stars through their recorded histories,
comparing their magnitudes to the surrounding stars in each case,
more than 3,000 in total.
And, accounting for differences in atmosphere, plate emulsion, telescope,
    and method of observation,
I have mathematically averaged to find an adopted magnitude for each.
It has, I will admit,
taken quite a lot of work.
(Oh, for the "computing machine" that we wished for at the turn of the
    century!)

From this we can find
the period and variability of each star,
something future scholars will certainly find useful.

And this is why I am most gratified
by Mr. Turner's generous words:
it is not simply the thrill
of discovering new stars.
It is providing "the care for their future."

American Woman, September 1907

I suppose I have complained enough
about the plight of women here,
the need for pay equal to a man
if her work be equal to his,
the need for recognition of the work of home
as important, and as consuming,
as her work at shops or offices or factories
or Observatories
here, in America.
But I have done important work,
raised a good son,
earned some measure of respect
in spite of being a woman and an immigrant
or, perhaps, because of it
here, in America.
I have spoken out
about women's work and capabilities
to young women students
and to the men who may hire them
and benefit from their work,
and hope to continue raising my voice
here, in America.
It cannot be long, surely,
until women are finally afforded
the right to attend colleges
(Harvard among them),
hold important offices,

vote in elections
here, in America.
And I have made my home,
my work,
my life
here, in America.
Now,
I am proud to say
that I have become a citizen
here, in America.

A Letter from Mr. Carnegie, 1907

I have continued correspondence with Mrs. Carnegie and her daughter.
Miss Margaret enjoys the small gifts I send her:
a pretty little box, a thimble, a moss garden in a glass jar,
a Bonnie Prince Charlie doll I sewed in Highland costume—
and she sends delightful letters.
Her mother sometimes writes as well,
reporting on the weather and hunting at their castle in Scotland.
Often she dictates to her private secretary
(a woman with beautiful penmanship),
so busy is she with her social engagements.

When I found a series of magic lantern pictures
of animals, wild and domestic,
I composed a rhyming story around them
and sent the scheme to Miss Margaret for her amusement.

Well,
who would have suspected that the man himself, Mr. Andrew Carnegie,
would read my words to little Margaret and her cousins,
enjoying the pictures cast in light upon the wall of their elegant hall
in the dark of a winter's night?

And, more,
who would suspect that Mr. Carnegie would send to me,
in his own hand
(though his handwriting be hard to decipher)

a letter of thanks and friendship?

"May á guid befá ye," he writes, wishing me well.
"Yours Ever, Andrew Carnegie."

I confess I'm proud to know the man,
and proud to be, like Mr. Carnegie,
an immigrant to America
and a Scot as well.

Medals, 1908

The Director tells me
he has renewed his nomination of me
for the Catherine Wolfe Bruce Medal for 1908.

However, the Astronomical Society of the Pacific
has instead awarded the Medal to Director Pickering himself.

Of course.
His "Revised Harvard Photometry," we hope,
will become the standard in the field,
the "rules" for measuring and photographing the stars
to be followed by astronomers the world over.
His insistence on developing a map of the entire sky
has significantly advanced the science.
The library of photographic plates
will no doubt prove useful to astronomers for years to come.
His diplomacy among those with competing theories
is well-known and appreciated by his peers.

His advocacy for women in Astronomy is admirable.
His hiring of them—of me—
is entirely appreciated.

Rest, 1909

The doctor,
summoned by my niece Mina,
insists I stay home for a fortnight
and recover my health,
neglected too long in favor of the work at the Observatory,
an all-too-short visit from my son,
the care of my mother and of Charles's wife Josephine
with her lying-in for the birth of their new little one.
(Och, I think sometimes how I would have loved
having another child, a daughter perhaps,
though how I could have managed I'm sure I do not know.)

I am cheered by a letter from Mrs. Louise Carnegie,
in Scotland for the summer where it has been cold
except, she says, for four summery days.
Nevertheless, the Carnegies have been
salmon fishing in the River Shin,
golfing, grouse hunting.
(Och, another daydream of visiting my homeland again,
though when will there ever be time for that?)

"It is all wrong," Mrs. Carnegie writes,
"that a woman who has done so much in the world
and whose life is as valuable
should . . . wear herself out.
We know the modest Scotch woman

who has already discovered more stars than anyone else
has still a lot more wonders to unveil for us."

In her kind words I take a warning:
that if I wish to continue discovering stars and caring for their histories
I must also care for myself.

The Works of Misses Maury and Leavitt, 1909

They are making names for themselves,
Antonia Maury and Henrietta Leavitt.

Miss Maury has been away from us
but not from Astronomy,
giving paid talks ($10 for a single lecture!)
featuring lantern slides of the stars,
using photographic plates that she has requested from the Observatory.
The Director and I have also, to encourage her research,
sent her our *Circulars, Annals,* and other news from here.
Now, ever restless, she returns to us for her work.

Miss Leavitt has been here,
studying stars in the Magellanic clouds,
until illness sends her to the hospital
and then to her family home.
We have sent her a viewing frame,
magnifier, flyspanker, journal,
and whatever photographic plates she requires
to assist in her pursuits:
she has found that the brighter stars have longer periods of variability,
though she does not know why.
Yet.

The Return of Mr. Halley's Comet, 1909

I had hoped, I suppose, to find it first,
but it is hardly the most pressing of my assignments.

Today there is a cable from the Royal Observatory in Greenwich:
Professor Wolf has seen it,
on photographs he took in Heidelberg on the morning of September 12th.

Given Wolf's measurements,
I make new calculations, consult our plates,
and find the faint impression—the blur
that is the light of the moving comet
impressing itself upon the photographic emulsion in a short line
while the stars in its background remain still, pinpricked points of light.

Now they have seen it through the great telescope
at the Yerkes Observatory in Wisconsin.
But it is still months until it will be visible to the naked eye,
months until it will pass close enough to Earth
to dispel the myths whipped up by the press
that the gases present in the comet's tail
will suffocate us all,
that disaster will befall—
or, perhaps, instead, great fortune.

The Director, speaking to the newspapers,
focuses on the scientific:

the revered astronomers and their calculations,
the illustration of the usefulness of photometry,
the distance and speed of the comet.

And he attempts to advertise, again,
the work of women at his Observatory:
"To Prof. Burnham, of the Yerkes Observatory,
belongs the credit of being the first man in America
to see the comet through a telescope,
and to Mrs. Williamina P. Fleming,
Curator of Astronomical Photographs at Harvard,
belongs the distinction
of being the first woman to see it."

Group Photograph at the Yerkes Observatory, 1909

The building is a palace,
all elaborate pillars and gargoyles,
marble floors and ornate ceilings.

Theirs is a large 40-inch refractor,
the same telescope that Mrs. Draper first saw
at the Columbian Exhibition in Chicago.

I am most interested in their Astrograph,
another telescope, like one of ours,
named for its benefactor Miss Bruce,
and designed for photographic plates 12 inches square.

The Astronomy and Astrophysics Society's committees have met,
papers delivered, luncheons consumed, press statements released.
(There is an effort this year to address the public frenzy
over the approach of Halley's comet,
as well as to tamp down the ridiculous calls to attempt to communicate
    with "Martians.")
I have presented my paper on "A Photographic Spectrum of a Meteor."
Miss Cannon and Miss Leavitt have presented also, on variable stars and
    standard photographic magnitudes.

We have traveled days on the train from Cambridge
(farther west than I have ever been,
and I will continue on my own,

days more to Salt Lake City to visit my son).
Other astronomers and physicists have traveled as well,
from laboratories and observatories around the world
to this Observatory on Lake Geneva.

Now we stand at the elegant entrance
looking out over Mr. Olmstead's lovely lawn,
assembled hastily by the photographer.
I again stand at the front—not due to reputation or fame,
but to provide a record of a woman here, among the men of science.
The few other women are sprinkled throughout the tiers,
distinctive in their white shirtwaists.
But I prefer my dark colors, more formal suits,
less discernible from the men in their dark suits, stiff collars and ties,
though I can unfortunately do nothing about my physical stature,
smaller than nearly every man present.

I find I am less patient with these photographs than I used to be,
sitting still, chin raised, in my father's studio so many years ago.
I am older now;
time is short,
and I have so much to do.

Farther West, August 1909

Such a relief it is to see the place my Edward has made his home.
He has written, visited, sent me postcards,
but it has been difficult to imagine him safe and happy
somewhere so far away.

I have taken one train west from Chicago,
then the Central Pacific across the high plains,
where the landscape seems nearly as endless as the sky,
to Salt Lake City.

Now I have met some of his friends,
seen his small office
(not surprisingly, rather a clutter
of papers and tools and samples of ore,
not unlike his pockets as a boy)
and a bit of his metallurgical laboratory,
where a staff works to find ways to join and separate metals from minerals.
And we are on our way to Zion Canyon,
where the landscape, he says, will take my breath away,
then to Los Angeles, where the streets are wide as rivers.

I see it all with my son.
Then I spend most of the long rail ride home
asleep in my berth.

Three Girls from New York,
December 1909

Their mother wished for them to visit
Radcliffe, Wellesley, Amherst, Tufts, Smith.
Harvard as well, though there is little hope
for the admission of women here
in time for even the youngest of the sisters.

I have shown them round the Observatory;
they have looked through the telescopes,
and in the evening played rummy and sipped tea and hot chocolate.

Their visit reminds me how much I enjoy having young people in the
    house.
Perhaps, if Edward can "settle down,"
there will be grandchildren visiting one day.

Och, at any rate,
it was pleasant to have young Nouvart, Vera, and Annie to stay.
And now, in thanks,
their mother Mrs. Costikyan has sent me a fine rug made on the family's
    looms,
a lovely and warm remembrance of their visit.

Another Decade, January 1910

Now, ten years into this young century,
and over half a century old myself,
comes a time to reflect upon my work.

I have now dozens of record books full
of my calculations, observations, speculations:
my reductions of stellar magnitudes,
my lists of peculiar spectra,
all the Draper work,
the research done for other astronomers
who have requested my verification of their theories and discoveries.

There are also my published results in the *Harvard Annals,* the *Circulars,*
the miscellaneous scientific publications,
the catalogues which I assembled myself
and those with which I assisted others.
There has always been, it seems,
a paper or table or catalogue of stars
requiring my revision, my approval, my guidance from pen to press.

I take a moment, while it is in my mind,
to note all of this in my current note book.
There is something about age and the new decade
that begs for a record.

My niece Mina will go out to a party tonight,

but I will stay in with my mother.
Charles and Josephine, with their little Josephine and the new baby
    Harold,
along with Charlie and Malcolm, nearly young men now, will join us.
Perhaps my brother Robert and his wife, as well,
or Johanna and her husband, will come,
if the snow ceases and the roads are suitable for the sleigh.

My brothers will debate the virtues of the automobile;
I will put on a magic lantern show;
Mother will rock the baby;
and we will raise a toast to "auld lang syne."

Collecting Stars, 1910

I suppose I must have seen them in the sky on moonless nights
when I was a girl in Dundee,
if the air in the city was not too filled with smoke.
But I cannot well remember it.

When I began my work here,
and read in the journals and catalogues the names of the stars—
and the constellations in which they reside,
I sometimes stood, on clear dark nights,
on a hill in Cambridge,
training my eye to find their patterns.

Then my eye was trained downward,
and I peered deep within the constellations on the photographic plates.
The stars there,
dark pinpricks in the negative white field, so faint,
and far more numerous than the naked eye could ever find.

There I found variable stars, novae, nebulae.
I have collected them, preserved them, cared for them.
Cared for them.

The Comet, May 1910

It is breathlessly reported in large, bold print
on the pages of the newspapers every day:
"COMET FACTS FOR TODAY"

but Mr. Halley's discovery has proved rather disappointing
to all but its most ardent admirers.

Though they have searched the skies to the west at the specified times,
the weather has not cooperated, the visibility poor.
The glimpses between the clouds have been dim,
not the fiery ball that many were hoping for,
though the poor comet has been blamed, somehow,
for storms, accidents, crimes, moods.

But we have a good photograph from Arequipa
and the Director has given it to the newspapers.
I suspect that, 75 years hence,
when some sage old citizen recalls from his youth
the sight of Halley's Comet,
it is likely that this view of it from the newspaper
is what he will remember as his own.

The Puzzle of Omicron^2 Eridani B, August 1910

I receive a note from the Director,
luncheoning in the Residence with Mr. Russell of Princeton.
He asks for the spectrum of a southern star, omicron^2 Eridani B.
I consult a number of plates on which the triple Eridanian star appears.
"Spectral type A," I reply,
and when the men appear in the Brick Building a bit later,
I learn of the cause for their request.

Henry Norris Russell has a collection of stars
of very faint magnitude, all of spectral type M.
Perhaps, he proposes, this is a rule, among similar stars?
My assertion of type A, which we suspect is associated with the hottest of
    stars,
seems to have spoiled his hypothesis.

"How, then," I ask, "can this star be both faint and hot?"

Mr. Russell is puzzled,
but Director Pickering is unperturbed.
"It is just such a discrepancy," he says,
"which will lead to the increase of our knowledge."

It is another puzzle
of which we leave the pieces
for our successors to assemble.

"The Best-laid Schemes," August 1910

We have been planning for months:
twenty foreign astronomers to shepherd through
the meeting of the Astronomical and Astrophysical Society here,
to the other nearby observatories,
on to Niagara Falls, Chicago,
The Lowell Observatory in Flagstaff,
then to the Grand Canyon,
to the International Solar Union in Pasadena,
and on to the Mt. Wilson Observatory in California.
Lectures, meetings, luncheons, dinners, field trips.
Two railroad cars reserved for collegial discussions along the way,
along with private compartments.

I plan to make the trip, assured that the women here—
one of them now my sister Johanna Mackie,
another my niece Ida May Stevens—
will keep up the work.
(Though August is when most take their yearly time off,
the rooms over-warm,
the days more suitable to boating and picnics,
the telescopes sometimes abandoned
due to hazy skies or thunderstorms.)
While the others go on after California,
I will travel to Salt Lake City to stay a week with Edward.
(He goes to South America soon, to study the copper in Chile.)

My niece Mina Davidson will keep up the house;
my mother, God love her,
will go now to live with the Mackies just over New Hampshire,
though she scarcely knows where she is these days.

I have done all the planning I can,
and though I have not felt my best since a bad cold this spring,
I'll not let it stop me from seeing the country
and from seeing the stars from all parts of it.

The Trip Begins in Cambridge, 19 August 1910

It reminds me, a wee bit, of the dock at Glasgow,
so many languages and accents speaking together
(though the language of scientists, fortunately,
is generally less coarse than that of longshoremen).
There are astronomers and physicists here
from Germany, Russia, France, Holland, Sweden, Switzerland, Austria,
  England, Scotland.

I am one of a small number of women at this conference,
though I am to be the only female delegate proceeding to the West.

There are lectures, technical sessions, committee meetings;
field trips to Blue Hill, to Sarah Whiting's Observatory at Wellesley,
to the student Astronomy laboratory in The Yard.

But this year, with such a congress of scientists here,
it is the communion among them that may prove most fruitful.
Director Pickering, President of this assembly,
has a particular intention:
to persuade the astronomers
to adopt his *Revised Harvard Photometry* as the standard for the field.
If he can accomplish an agreement on that, he says,
he will then begin a campaign for the adoption
of one system of stellar classification to be used in all of Astronomy.
Ever the diplomat,
he claims not to have a preference among the three or four systems in use.

He says, however, that the Draper classification system,
devised by me under his direction,
will surely be one of the favorites.

The last sessions in Cambridge finished,
we assemble trunks and valises,
distribute ourselves amongst the waiting carriages,
and depart for South Station
from whence we will begin our Astronomical Tour of America.

Niagara, August 1910

The Falls are immense, powerful, thunderous.
Carriages take us to an island near the crest of the American Falls
where I brave a turn at the railing to watch the river
rush headlong over the edge.

Later, below the Falls we board a boat
and steam so close to the spray that we are sodden and shivering
the rest of the afternoon, despite the August heat.

Our foreign visitors are quieted
by the roar, and the sight, of the Falls.
But in our carriages, at our meals,
in the rail cars next day as the country rushes by
the discussion is all about
the comet, the Sun, the stars.

Chicago, August 1910

I would like to have been at the World's Fair,
to have seen the White City,
the feats of engineering and electricity,
the Women's Pavilion,
the art and the architecture,
and to have read my paper aloud here, back in 1893.

But I am here now, in 1910—
not a computer with a modest proposal
for "A Field for Women's Work in Astronomy,"
but a colleague, a "fellow"
amidst the world's foremost astronomers.

True, I am often assumed, at first,
to be one of the wives accompanying a scientist husband,
assumed to have little but pleasantries to contribute to the conversation.
But my name precedes me.
They have seen it on my publications,
on the letters from the Harvard Plate Collection
verifying—or denying—their hypotheses,
on the tabled star catalogue listings as "discoverer."

"Ah, so this is Mrs. Fleming!" they say, upon my introduction,
and then include me, however briefly, in their circles.
Even in this "enlightened age,"
I cannot expect to be treated as an equal.

I have not had the opportunity, as they have,
to devote myself to study,
as this visit, to the University of Chicago, reminds me.
But I have learned.
Through work, I have learned.

And though by title I am Curator,
by work I am also Astronomer.

Night Train, August 1910

We board the train on Monday evening
and although discussions continue on in the astronomers' cars,
I retire to my compartment,
remove my shoes, loosen my clothing,
sit upon the narrow bed—
and awaken, still dressed, in the middle of the night,
somewhere in the middle of America.

I can see nothing but the stars
outside my compartment window.
But now that I am finally properly abed,
they are all I need.

On to Arizona, August 1910

We are on the train nearly all of Tuesday and Wednesday,
the 23rd and 24th of August.
The rail cars, it seems, are ovens,
each breath drawn in from the open windows
as hot as the last.

The men have forsaken their jackets,
rolled up their shirtsleeves, loosened their ties.
Those who have not been cajoled by the Director
to join him in discussions of photometry
nod off in their seats.

I am as sociable as I can bear to be,
appearing for meals and brief conversation,
a walk along the platform at the longer station stops.
I retreat to my compartment
between my shows of collegiality,
stifled by my high-necked, long-sleeved dress and petticoats,
my corset, stockings, and shoes.
I remove what I decently can,
loosen the rest, close my eyes,
and try to remember Niagara Falls.

Flagstaff, August 1910

We arrive late Wednesday,
and though we are all weary,
the Lowells insist on a party.
Their bungalow, fortunately,
is cooler than the train,
and Mr. and Mrs. Lowell are charming and enthusiastic hosts,
entertaining us until the wee hours.

In the morning we visit the Lowell Observatory,
where, they say, the climate is ideal for studying the skies,
rarely enclouded and nearly arid.
While Percival Lowell discusses his Mars studies
(his idea of the "canals" there built by some "Martian" beings, thankfully
   now abandoned),
Director Pickering, standing next to me for a moment,
rocks back and forth on his heels.
"I've done it, Mrs. Fleming," he says.
"They have agreed.
My Harvard Photometry system will be the world standard for assessing
   magnitudes."
He leans toward me to make sure I am able to hear.
"It is likely the most important thing I have done," he says.
Then he turn to join another conversation,
this one regarding the effects of the recent comet.

Now, I know, the Director will begin another quest:

one worldwide standard for classifying stars.
The system I developed for organizing the Draper Catalogue
will be among the candidates.

The View, August 1910

Next morning, we take in the Grand Canyon
to the audible gasps of some of our European colleagues—
a walk along the rim and out to a point affording magnificent views.

Here, I wonder:
if one could stand on Mars—or nearer, on the Moon—
what would one see of the Earth?
Is the red rock of this Canyon visible there?
The green of the forests?
Or is it, I suppose,
the blue of the oceans
that would distinguish this planet from each of the others?
We look so often upon the stars,
our Sun and Moon,
the other known planets,
but we do not yet know how our own would appear
through the eye of a telescope.

Farther, and Farther, West, 1910

Saturday evening we board the train once more,
west across the Mojave Desert.
I write letters
as the darkness envelops the train,
and as the temperature becomes surprisingly cooler inside the
compartment,
I look out at the dark landscape
lit only by a halfmoon,
stars all the way to the low horizon.

Pasadena, 1910

After the sparse desert landscape
dotted occasionally with groves of trees
unlike any I have seen before,
cacti of various, curious shapes and sizes
(particularly interesting to those members
with botanical interests in addition to the astrophysical),
and heat that rises nearly to 100 degrees F,
Pasadena is an oasis.

We arrive late Sunday afternoon
and find our rooms in the Maryland Hotel,
nearly completely occupied by our party.

We are now the guests of George Ellery Hale,
his Mt. Wilson Observatory, and the Solar Union.
It was Mr. Hale's invitation that began this journey,
a suggestion that Director Pickering's influence
and the lure of an American "Grand Tour"
might bring together scientists studying stars
in addition to our Sun.

At the Pasadena offices of the Mt. Wilson Observatory Monday morning,
I meet the staff of computers, primarily women,
doing work very similar to that of my staff at the Harvard College
    Observatory.
(Here they count Sun spots rather than stars.)

I inquire about their training, their procedures, their record-keeping,
and, more discreetly, their hours and their pay.

I am immensely pleased that Harvard's example
of women at work in Astronomy
has been copied, now, by other Observatories.
I am dismayed, however,
that the women so employed
are still paid less than the men,
in Cambridge, in Pasadena, everywhere,
their "girl-hours" of work computed into departmental budgets
at a fraction of the cost of the work hours of men.

My new friends here, though,
are rightly proud of their work,
and as we tour the rest of the offices,
and attend a garden party at the Hales',
it is comforting,
after a week in the company of male astronomers,
to be among "sisters" again.

Up the Mountain, August 1910

On Tuesday we board a trolley
from Pasadena to the foot of the mountain
which rises rather abruptly from the surroundings—
no rolling hills here—
and then we choose our conveyances
for the trip up Mt. Wilson to the Observatory.
The younger and more adventurous sit astride horses or burros
and others decide to walk, but I and the remaining scientists and computers
take seats in open carriages pulled by teams of horses.

The trail rises above the valley,
and the view below is of long rows of vines laden with grapes
and groves of trees that I'm told will bear oranges.
There is a camp halfway up the mountain,
where we stop in the shade
and, seated upon log benches or ledges of stone, we eat boxed lunches.
After a brief rest for men and horses,
we go upward, the trail so steep and winding, the turns so sharp
that several times we must disembark from the carriages
lest a wheel stray off of the curve.

At the summit we are mercifully assigned to cottages,
primitive but comfortable.
We are rewarded, at the end of a long and warm and dusty day,
with pitchers of water, clean cots, a cool mountain breeze,

and a view of the tiny sparkling electrical lights of the city in the distance
    below
and the constant light of the stars in the distance above.

Mt. Wilson, August 1910

Atop the mountain, in the mornings,
there are examinations of the telescopes,
committee meetings, sessions of general interest—
all conducted in English, German, French.

Mr. Hale, of course,
presents his spectroheliograph, discusses the Sun,
and presses for the Solar Union to add other stars to its study.
Professor Kapteyn discusses star streams—
stars moving in the same direction and at the same rate,
though he has no explanation, yet, for the phenomenon.
Director Pickering presents the case for collecting opinions
on the systems of classifying stellar spectra,
with the aim of accepting one unifying standard.

There are "siestas" in the heat of the afternoons,
for those not coaxed into debate.
(Some of the wooden buildings here
are ingeniously outfitted with capes of white canvas,
reflecting the sunlight and creating shade.)

Late in the evenings, through the telescopes:
the Hercules Cluster, the Ring Nebula and variable stars,
Saturn.

The Case for Harvard, August 1910

The Germans favor Vogel's system,
based on heating and cooling stars,
though the discovery of helium has necessitated a revision.

Several astronomers favor Miss Antonia Maury's system
built upon hydrogen absorption lines and added criteria,
a complex system that is more descriptive,
but unwieldy and unyielding to new discoveries.

The "Harvard System," the lettered classes
which I devised for the Draper catalogue
on the basis of the strength and pattern of spectral absorption lines
—arranged by Miss Annie Jump Cannon into a new order—
is another candidate.
Though it is the work of two "women computers,"
it carries the name of our revered institution
and its respected Director, Pickering.
Though its order is discordant,
it has been adopted by many astronomers.
And although it is less intricate,
it has already accommodated new discovery.

Several in attendance speak out in Harvard's favor.
No decision will be made here.
A survey will be constructed,
responses gathered from a questionnaire.

But it appears that our system—
Draper's, mine, Cannon's, Pickering's, Harvard's—
may yet prevail.

Descent, September 1910

We rise early on Saturday,
collect our belongings,
and begin our precarious descent.
The carriage driver must rein in the horses,
anxious to relieve their burden and reach the valley below.
Where the path is too narrow or sharply curved
we must again disembark,
some (not I) venturing close to the edge
to investigate the peril.

Finally, dusty and weary, we return to our hotel.
In my room I pour water from the pitcher to the basin,
wring out a wash cloth,
unpin my hair,
and bathe myself from head to toe.
I should dress for dinner,
a final banquet hosted by the Hales.
I delay, taking the time
to write to my mother and Johanna of my adventures—
the travel, the heat, the mountains,
the astronomers I have met,
their support for my Draper classification system.

I will post the letter tomorrow before I leave on my own.
The astronomers will be dispersing,
some boarding a train to the Lick Observatory,

a night and a day's ride north of here,
then by rail back to the east coast.
Others will visit friends, colleagues, family
before returning to their home countries.

I will take the train to Salt Lake City
for a week with my son
before returning to Cambridge.
I confess
I have looked forward to this time with Edward
more than any other portion of the trip.

Banquet, September 1910

We assemble one last time in the hotel ballroom,
more than eighty astronomers and astrophysicists
from fifty observatories in thirteen countries,
local dignitaries and patrons of Mt. Wilson Observatory.
It is an elegant affair,
our hosts gracious and the company both charming and interesting,
all looking quite more presentable than we did just a few hours ago.

Cablegrams have been read aloud
from Lady Huggins in England
(recently widowed, she and her husband worked tirelessly in spectroscopy)
and from Mr. Andrew Carnegie
(a major patron of this Observatory, though his contributions to our own
    have diminished),
congratulating the assembly on its work.
Resolutions are read,
organizers congratulated,
awards presented.

I enjoy the meal and the company,
but I fear the journey has caught up with me.
I retire early
and sleep, I think, without dreaming.

Dinner with Edward, September 1910

My train arrives at Salt Lake City late Monday.
Edward has borrowed an automobile
to meet me at the station, and take me to dinner at his club.

He is attempting to impress me with his success here, I am sure.
But it is not the automobile, nor the club, which pleases me,
but the enthusiastic talk of his work,
the way he is looking forward to the future.
There is to be a posting in South America—
Chile, he expects, to work in copper.

There is no mention of a sweetheart,
but I will investigate that
in the next few days of my stay.

Telegram, September 1910

But
when we arrive at his rooms,
a telegram awaits us,
slipped beneath his door.

He frowns and opens it,
reads and goes pale.
"It's Grannie," he says, and hands it to me.

It is from my sister Johanna.
Our mother, Edward's grandmother, has died.

Taking Leave, September 1910

Edward drives us to the station,
our visit cut short just as it begins.
He offers to accompany me back to Cambridge,
but I insist he stay and tend to his work.

He offers me his arm as we wait on the platform,
and I hold it tight.

"We must remember," I tell him,
"to treat each 'goodbye' as if it is our last.
We dinnae ever know
when we may see each other again."

"Aye," he says, handing me his handkerchief.

"I am so proud of you, Edward," I say
as the train pulls in.

"Mam," he says,
"it is I who should be proud of you."

Remembrance, September 1910

I awake in the dark.
The steady rocking of the train
has somehow combined with my weariness
and the valise in my lap
to convince me, dreaming, that I have been rocking a baby.

Och, my mother.

She is gone now.
And though for months she had been fading away from us all,
it is still a shock to wake and rediscover it.

She rocked babies.
Each bairn of her own, so many,
my sisters and brothers and I,
each scarcely walking
before the next arrived to take his place.
She rocked them to nurse, to sleep, to comfort,
and though most grew fine and strong,
there were others she rocked,
born too early, or sick, or weak,
whom she rocked though she knew they would not last.

And then she rocked her grandbabies,
began their raising,
while we began to make a life for them.

She kept a house and food on the table,
aided her own mother in her old age,
helped my father with his shop of picture frames and wooden boxes and
    photographs
and, I think now,
she helped to keep the creditors at bay.

And now that she is gone,
I wonder, while she tended to her home, her husband, her babies,
what other talents of hers lay undiscovered?

Och,
had the opportunity been hers,
had she wanted it,
what else might my mother have done?

A Generation, September 1910

We gather after the service,
my three brothers, my sister, and I,
remembering our dear mother,
our childhoods, our father, our siblings who did not survive their own
    childhoods,
one who remained in Dundee, all these years,
all gone.

As the day goes on
and the kind neighbors take their leave
and the pastor returns to his church,
the tears give way
to smiles and memories.

It is we who remain of the Stevens family,
now the elder generation,
together, here, three thousand miles from where we began.

We marvel at the grey in each other's hair,
compliment the accomplishments of our nieces and nephews
—most now grown—
and tease Charles, the youngest of us,
with a new wee one on his knee.

It is thirty-one years since I was in Dundee,
without a husband, a bairn just weeks away.

It was my mother,
my brother Charles not yet a man,
and my dear Grannie
who helped me then.
And when I returned to Cambridge,
there soon were brothers and a sister here
so I was not so all alone.

Today, our mother gone,
we are fortified to know
we have each other.

Back to Work, September 1910

Within a fortnight, the Observatory is back to work,
the computers who had taken their August leave,
the Director from Mt. Wilson,
myself from my bereavement,
all returned.

Director Pickering is triumphant.
Having secured his Harvard Photometry as the standard in the field,
he left the Solar Union with approval for his next mission:
to determine the views of major astronomers
and to recommend a standard stellar classification.
The surveys he has sent out are already returning,
and the Harvard-Draper system—our system, my system—
is receiving most favorable reviews.

I have tables to assemble for my next catalogue of stars,
the *Spectra and Photographic Magnitudes of Stars in Standard Regions*,
dividing the sky into forty-eight 5-degree squares,
displaying the location, magnitudes, and spectral class
of the stars residing in each.
I also continue the work on stars with peculiar spectra.

And the work I most prefer:
I study the plates for whatever they may reveal to me.

Discoveries, October 1910

My recent searches have been rewarded:

New Star in Sagittarius.— *A new star whose approximate position is R. A. 17h 52m 15s, Dec. -27° 32'.3, (1875), was discovered by Mrs. Fleming in the Constellation Sagittarius on October 1, 1910. It appears on sixteen photographs taken at Arequipa with the 8-inch Bache and 1-inch Cooke Telescopes, between March 21, 1910 and June 10, 1910. The magnitude has been estimated as varying from 7.8 to 8.6, between these dates. The spectrum is quite faint but shows the bright hydrogen lines Hβ, Hγ, Hδ, Hε, Hζ, and Hη, with a trace of Hγ as dark on the edge of greater wavelength of the bright line Hγ. The star does not appear on seventeen photographs, taken between July 23, 1889 and October 7, 1909, although most of them show stars fainter than the twelfth magnitude and one plate shows stars of the fifteenth magnitude, or fainter. An observation by Leon Campbell on October 3, 1910, with the 24-inch Reflector of this Observatory confirms the presence of this object and gives its magnitude as about 10.5. Of the fifteen new stars knows to have appeared during the last twenty-five years, eleven have been found at this Observatory, nine by Mrs. Fleming from the photographs of The Henry Draper Memorial.*

E. C. Pickering, Harvard College Observatory, Cambridge, Mass. Astronomical Bulletin No. 426.

Nova Arae.— *A new star whose approximate position is R. A. 16h 31m 4s Dec. -52° 10′.6 (1875) was discovered by Mrs. Fleming, in the constellation Arae on October 13, 1910. It appears on 21 photographs taken at Arequipa with the 8-inch Bache and 1-inch Cooke telescopes, between April 4, 1910 and August 3, 1910. The magnitude has been estimated as varying from 6.0 to 10.0 between these dates. The spectrum is quite faint but shows, on three plates, the bright lines, 5007, Hβ, 4670, Hγ, Hδ, Hε and Hζ, one of the plates showing also the bright line Hη. Apparently this object had passed into a nebulous condition before its spectrum was photographed. The star does not appear on 44 photographs, taken between August 20, 1889 and March 19, 1910, although almost all of them show stars fainter than the twelfth magnitude, and two plates show stars as faint as the fifteenth magnitude.*

Of the sixteen new stars found during the last twenty-five years thirteen have been found at this Observatory, one by Miss A. J. Cannon, two by Miss H. S. Leavitt from photographic charts and ten by Mrs. Fleming from the Draper Memorial photographs.

Edward C. Pickering.   Harvard Astronomical Bulletin, No. 427.

To see something that has never been seen before—

it is as if I am an explorer
in a sea of stars
discovering a new world.

Headlines, 1910

I suppose it is the recent comet
that has spurred interest in Astronomy.
Or perhaps it is the righteous argument for women's suffrage
that has made a woman in science so noteworthy.
At any rate, the newspapers have discovered me again.

From the "Pittsburgh Post-Gazette":

> *NEW STAR DISCOVERED—Mrs. Williamina Fleming
> Adds to Her Astronomical Record*

and the "Pittsburgh Daily Post":

> *ANOTHER NEW STAR IS ADDED TO WOMAN'S
> LIST—Mrs. Williamina Fleming, of Harvard Observatory,
> Makes Discovery in Firmament*

"The Baltimore Sun":

> *WOMEN IN ASTRONOMY—Fair Sex Has Made A
> Remarkable Record in Study of Stars*

the "Fort Worth Star-Telegram":

NEW STAR DISCOVERED BY WOMAN ASTRONOMER—Sixteen Have Been Found in Last Twenty-five Years and Men Have Found but Three

and I must send the last to Edward,
from "The Oregon Daily Journal":

THE WOMAN WHO FOUND FAME IN THE HEAVENS—The Untiring Work of Mrs. Williamina Fleming, Who Has Discovered Ten New Stars by Scanning Thousands of Solar Photographs—"the Sherlock Holmes of the heavens"

"The great detective!" Johanna exclaims.
"I suppose this implies that one of us is Dr. Watson?" asks Miss Cannon.
I am happy to accept their gibes.
The attention is ridiculous.
But this—
detecting a point that I do not recall on a photographic plate,
and looking for it, backwards,
through ten, twenty, thirty years of photographs,
until I am satisfied that no one has ever seen it before—
this is the work
that I would do
to the end of my days.

A Surprise Party, December 1910

It is a particularly busy day in December,
and I am to and fro between my office, the computing room, and the stacks
    of the plate collection,
much requiring my attention.

I take a call in my office,
and upon my return to the computing room find the work cleared away,
trays of cake and tea and little squares of fudge upon the tables,
and the women standing about expectedly.

"What is all this, then?" I exclaim.
There is work to do, and our Christmas luncheon is not for another week.

Just then the Director enters through the other door,
and the Reverend Metcalf—a friend and pastor as well as amateur
    astronomer
who had telephoned me after ten o'clock on an evening last August
to tell me he'd seen a comet—accompanying him.

The Director stands, rather formally, and clears his throat.
"Mrs. Fleming," he says,
"yesterday morning, I received a cable from the American Embassy in
    Mexico.
The Astronomical Society of Mexico has awarded you a medal."

Reverend Metcalf interjects.

"La Sociedad Astronómica de México," he pronounces carefully.

"Yes," the Director says, consulting a note in his hand.
"The Guadalupe Almendaro Medal for the discovery of new stars."

There is a polite round of applause.
"Brava!" Miss Maury exclaims.

"The medal itself is in the custody of the Deputy American Ambassador,
but it shall make its way here, eventually," the Director says.
"It is quite an honor, Mrs. Fleming."

"Hear, hear!" says Miss Cannon,
and the others join in as Miss Winlock pours tea.

"And I have received a medal, as well,
thanks to your help, Mrs. Fleming," the Reverend Metcalf says proudly.
I was pleased to study the photographic plates
to find that he had, indeed,
been the first to note his comet last summer.
He has received the Sociedad's Felipe Rivera medal
for the discovery of comets and asteroids.
He raises his teacup, and I notice that Mrs. Metcalf is present, as well.
"How do we say it, my dear?" he asks her.

"'Slàinte mhath,'" she answers, smiling at me.
"To your health, Mrs. Fleming!"

The room erupts again.
My Observatory "family,"
who have gently teased me for my Scots accent all these years,
exclaim with a fair pronunciation of the Scots-Gaelic cheer,
"Slàinte mhath!"

Over tea and cake, I ask
how they ever managed to arrange the celebration

without my knowing.

"Mrs. Fleming," Miss Cannon says, triumphantly,
"you cannot supervise everything!"

Working with Miss Cannon, January 1911

Miss Cannon is conducting a bit of an experiment.
Our spectrographs, improved by more sensitive photographic plates,
now reveal a more detailed spectrum of each star.
Will our classification system, she wonders,
crumble under the new information,
jumbling our classifications, or will it hold fast?

To test, she is studying new photography—
that better catches the lines of the "red" end of the spectrum—
of the stars which we have already classified.
(She is particularly adept at reading spectra,
identifying at a glance what others require a magnifier
and considerably more time to decipher.)

Her examination leads her to some minor adjustments,
but she triumphantly reports that the system that I began
as an organization of lettered groups of spectra,
set into an order by Annie Cannon herself, holds!

"This," she says, "will further support our system—
the Pickering system, the Draper system, the Harvard system,
whatever we are to call it—
as the prevailing standard for classifying stars."
She muses.
"And what <u>should</u> it be called?"

"Such things are usually named for the director
under whom they were developed," I say,
"or the source of its funding.
so 'Pickering,' or 'Harvard,' or 'Draper,' by convention, I suppose."

"Mrs. Fleming," Miss Cannon says,
and we lean even more closely together
than we have done to aid her hearing,
"let us be unconventional just for today."

I look up.
We are astronomers, Miss Cannon and I, after all.
I think we have proved ourselves beyond computing.
Why should we not claim it?
"The 'Cannon-Fleming System' it is, then," I laugh, "for today!"

To the Printer, February 1911

I have delivered another *Harvard Annals* to the printer,
with my most recent catalogue of stars in standard regions.
Now, until I have the printer's proof in hand,
perhaps I can return to the plates.

I found only eight new variable stars in the last year,
so busy was I with other work.

I fear that if I slow down,
exhaustion will overtake me,
as tired and weak as I have been.
There is ever so much to do.

But first, a rest.

The Work Goes On,
February-March 1911

I return in February
from a few days at home, nursing a bad cold,
to a new variable star discovered by Miss Leavitt,
and examine a series of plates
to determine to which class it belongs.
Then there is an examination of plates for a new variable
identified by Mr. T. H. Astbury, the English country schoolmaster
who previously managed to discover a Cepheid variable with only his
    naked eye.
I am pleased to verify that he has indeed found another.

But I find I accomplish less each day.
Is it age, I wonder?
Though I am "only" fifty-three,
I have never found it so difficult
to do as much—or more—than is asked of me,
or that I expect of myself.

Heeding Advice, March 1911

Perhaps I should finally take the advice
of Mrs. Draper and Mrs. Carnegie,
and take some time away for rest or travel or adventure.

I have met Mr. Charles Glidden, the aeronaut, on several occasions,
and have decided to take one of his gas balloon trips this summer,
"floating" from Boston to New York!

And I hope, next year—in August, perhaps,
when the seeing is not as good at the telescopes—
to return to Scotland for several weeks.
Perhaps the Carnegies will welcome me to their castle in Sutherland.

Perhaps also, in the next few years,
I will accompany Miss Cannon,
who has expressed the hope of using the telescopes in Peru,
an adventure that would place me nearer to Chile and my son.

But now, again,
what I appear to need most is rest.

Thirty Years, 9 May 1911

For thirty years,
once I left the servant's work of the household
and found my place here in the work of the Observatory,
I have calculated the reductions,
estimated the magnitudes,
examined the spectra,
classified the observations,
catalogued the stars.

I have supervised, in my estimation, more than forty women computers.
some have done no more than the mathematical calculations,
others the assessment of magnitudes or the examination of spectra.
Others have become investigators.
They have discovered new stars;
they have observed and wondered and theorized and tested
and often proved their hypotheses correct.

There are new photographs, new understandings.
But I also return to the historic collection
to examine anew what was once thought to be error, anomaly,
and might now, with a modern interpretation, be better understood.
I have spent the last few days, when I've had the strength,
with those photographic plates
and my notebooks full of remarks on the stars.

But today I have found it difficult to concentrate on my work.

It is hard to catch my breath,
and as I pass my hand across my brow
I fear I may be feverish.
I ask Miss Leland to call for a carriage,
ask my sister Mrs. Mackie to collect things for me from home,
and replace my notebooks carefully on the shelves.
I ask the carriage driver
to deliver me to the hospital.

A Difficult Time, May 1911

It is worse than I thought.
Pneumonia—
a relentless, painful cough,
fever, fatigue, and weakness
to which I am entirely unaccustomed.

Johanna has brought me my papers,
but I find it difficult to work on correspondence.
I cannot concentrate enough to read anything of substance,
and can hardly hold a pen steady enough to sign my name.

"Rest," the doctors say,
and though it is against my nature
to surrender,
I close my tired eyes.

In Dreams, May 1911

When I have awakened,
Miss Cannon has been here, Miss Leland and some of the others,
Reverend Metcalf with Director Pickering.
My sister Johanna has been often at my side,
and my brother Charles as well, I believe.

But I have also imagined Edward here,
though I know he is in South America,
and my mother here, and my Grannie, here from Dundee.
Though that cannae be correct, can it?

I am hot and cold.
I am in my hospital bed,
in a chair by the stove,
on a steamer in the Atlantic,
on a train across the country.

And now I hear my sister's voice,
though I cannae follow what she says,
and then my brother's.
"We had best send for Edward," he says.

Into the Ether, 21 May 1911

I cannae seem to escape the fever
and the pain in my chest
and the tortured sound of my own shallow breathing

but as I gaze at the ceiling of the hospital ward
I can imagine an escape from this sickness
and from this narrow bed.

The white painted ceiling gives way
to the cool dark night sky
and I float up through the stars,
each flaming warmly as I draw past.

Ahead is the triangular wisp of a nebula,
a veil becoming more distinct as I approach it—
a cloud of dust and stars and more,
taking shape and color,
beautiful color—

and as I draw nearer and nearer
I know that to discover
what this nebula—
this lovely
nebula—
is made of,
I need only

to . . .

reach . . .

out . . .

Epilogue

Williamina Paton Stevens Fleming died of pneumonia on May 21, 1911, two weeks after being admitted to a Boston hospital. She was 54 years old.

Her grave marker in Mount Auburn Cemetery in Cambridge, Massachusetts, reads "Williamina Paton Fleming / Astronomer."

Mina Fleming discovered 10 stella novae, 310 new variable stars, 59 nebulae, and 94 Wolf-Rayet stars.

She classified the spectra of over 10,000 stars.

She organized and cared for over 200,000 glass photographic plates in the Harvard College Observatory collection.

She was the first person to notice what later became known as the Horsehead Nebula and the first to notice the part of the Veil Nebula now known as Fleming's Triangular Wisp.

She was the first to recognize the existence of the stars later called White Dwarfs.

She was the first woman to hold a title at Harvard College, the Curator of Astronomical Photographs.

She supervised over 40 Harvard women computers, many of whom made their own important discoveries and greatly advanced the field of astronomy. Among them are Annie Jump Cannon, Antonia Maury, and Henrietta Leavitt.

She advocated for better pay and recognition for the Harvard women computers and all working women, and supported women's fight for voting rights.

She was named a fellow at Wellesley and Radcliff Colleges, despite having the equivalent of only a high school education.

She was named an honorary fellow of the Royal Astronomical Society, the first woman working in America to be so honored.

She was awarded the Astronomical Society of Mexico's Guadalupe Almendaro Medal of Astronomy.

Despite numerous nominations, she was never awarded the Bruce Medal in Astronomy. In fact, no woman was awarded this medal until 1984.

Meteor 5747 Williamina, discovered in 1991 by R. H. McNaught, was named in honor of her.

In 1970, The Fleming Crater, a feature on Earth's Moon, was named by the International Astronomical Union to honor both Nobel laureate Alexander Fleming and Williamina Paton Fleming.

In his remembrance of her for the Royal Astronomical Society in 1911, British astronomer Herbert Hall Turner wrote:

> "As an astronomer Mrs. Fleming was somewhat exceptional in being a woman; and in putting her work alongside that of others, it would be unjust not to remember that she left her heavy daily labours at the observatory to undertake on her

return home those household cares of which a man usually expects to be relieved. She was fully equal to the double task, as those who have had the good fortune to be her guests can testify : and it is perhaps worthy of record, as indicating how lightly the double burden sat on her, that she yielded to none in her enjoyment of a football match, especially a match between Harvard and Yale."

Remarks on Fact and Fiction in this Imagined Memoir

1—Cambridge, 1905

Mina spoke with a Scottish accent, which I've represented with a few words and phrases throughout the story, more pronounced early in her career and later when she is with family or in distress. Her writing voice and her unique personality I based primarily on her Harvard Journal of 1900, the few letters which are collected at Harvard, her published writings, and quotations from her in newspaper articles of the time. Direct quotations from Mina and others are represented in italics throughout the story.

Mina was first nominated for the Catherine Wolfe Bruce Medal in 1900 and again in several subsequent years.

2—Boston, 1879

Mina (21) and her husband of 19 months, James Orr Fleming (37), arrived in New York City December 3, 1878 via ship from Glasgow, and then settled temporarily in Boston. He had worked for a bank in Dundee.

3-Morning, 1879
4-Discovery, 1879
5-One Day of Despair, 1879
6-A Day of Determination, 1879

These scenes are largely imagined. Mina did not describe the circumstances, but did describe the separation from her husband as abandonment.

7—Beginning Again, 1879

Although she did not describe the circumstances of her abandonment,
Mina seems to have implied later in interview that she and her husband
were aware of her pregnancy before he left her, and that she felt desperate
to support herself.

8—Climbing Observatory Hill, 1879:

Mina apparently shared the story of her grandfather's connection to the
"Fighting Grahams" of The Highlands with friends and colleagues, and
it is consistent with records in her family tree.

9—A Roof, 1879

Edward Charles Pickering (1846-1919) and Elizabeth Wadsworth Sparks
Pickering (1849-1906) occupied the Director's Residence, which
was attached to the Harvard College Observatory. The Residence's
household finances were the responsibility of the Director.

10—A Meal, 1879

Mina was hired as "second maid" sometime in the spring of 1879. Although
this scene is imagined, it was quickly apparent to the Pickerings that
Mina was quite intellectually capable. It is also apparent through several
of her writings that Mina was fond of fudge and prepared it often at
home.

11—Observatory, 1879

Parts of the Harvard College Observatory buildings remain today on
Observatory Hill, and are used by the Harvard University Department
of Astronomy. They now form part of the Harvard-Smithsonian Center
for Astrophysics.

12—Novae, 1879

A **nova**, in Mina's time, was a sudden, bright star, seen for the first time.
Today, **novae** (plural), stars which explosively shed their outer layers but
are not destroyed, are distinguished from **supernovae**, which are stars
violently exploding and extinguishing.
A **comet** is a ball of ice and debris which orbits the sun. Comets are visible
when their elliptical orbits bring them close enough to Earth.

A **refractor** is a telescope that uses a lens to gather light. A **meridian circle** is a telescope mounted to move only along a north-south line, to observe objects as they cross this meridian.

The **Firth of Tay** is the estuary and coastal waters into which the River Tay of Scotland empties, where the city of Dundee is situated.

The Latin Scholar (as are the other few characters without surnames throughout the story) is imagined; however, the Pickerings frequently entertained guests from the College, Academia, and Society.

13—Daguerreotype, 1879

Louis Daguerre, a French artist, is credited with inventing a process using light and chemicals to produce a remarkably accurate image. His secret process was purchased by the French government and published "free to the world." Mina's father, Robert Stevens, was one of the first in Scotland to learn and use the process commercially.

14—Alterations, 1879

Mina was an accomplished seamstress. She sewed and sold costumed dolls for gifts and charities, and was described in the 1881 Scotland census as a "dressmaker."

15—Society, 1879

Lizzie Pickering, the daughter of a previous president of Harvard, historian Jared Sparks, was a member of Boston society with a reputation as a gracious hostess.

Henry Wadsworth Longfellow (1807-1882) was living in Cambridge at this time, and was nationally famous for his poems "Paul Revere's Ride," "Song of Hiawatha," and "Evangeline."

Robert Burns (1759-1796) is still revered as the national poet of Scotland for his poems such as "Auld Lang Syne," "To a Mouse," "Tam O'Shanter," and "A Red, Red Rose."

16—The Garden, 1879

Although this is an imagined conversation, Mrs. Pickering was an avid gardener and did command the grounds of the Observatory. Her husband did, in fact, sell the grass clippings.

The Pickerings did not have children.

The details of Mina's childhood are taken from interviews granted local and national newspapers.

17—Useful, 1879

This scene is based on an often-reported story of how the "Scottish maid" came to do the work of male assistants at the Observatory.

18—Transit, 1879

In astronomy, **transit** is when a planet appears from our view to cross the face of the sun, or a moon to cross its planet. It can also refer to a celestial body's path across the meridian. Here it serves as a metaphor for Mina's path from maid to computer.

19—In the East Computing Room, 1879

Mrs. Rebecca Titworth Rogers worked at HCO from 1875-1898, assisting her husband, astronomer William Augustus Rogers, with his calculations. Miss Anna Winlock worked there from 1875-1903, starting soon after the death of her father, the third director of HCO. Miss Selina Cranch Bond (daughter of the first HCO director, sister of the second) worked there from 1879-1897. Miss Rhoda G. Saunders worked there from 1875-1898. Although she had no prior connection to the Observatory, she appears to have been recommended for the position by then-president of Harvard, Charles Eliot.

20—New Work, 1879

Although she had become a computer, Mina was probably paid from the Pickering Residence's account into 1881.

21—Computation, 1879

The astronomers usually provided the formulae for the women computers to apply to the data from their nightly observations, based on a variety of factors; their resulting reductions were recorded, sometimes averaged, and catalogued, Of course, at this time, before the existence of modern computing machines and calculators, the people doing the math were called computers.

22—Respect, 1879

Mina's organizational and administrative skills, in addition to her computing, were apparently quickly recognized and appreciated by Director Pickering and the other women.

23—An Imprecise Calculation, 1879

By the summer of 1879, Mina would have been noticeably pregnant; the estimation of due dates was quite inexact.

24—A Word, 1879

This conversation is imagined. However, a similar conversation must have taken place, as Mina did travel to Scotland with the expectation of returning to HCO, and was consulted about the furnishing and design of the computing room. That the Pickerings probably paid, or at least assisted, with her travel expenses, is an assumption.

25—The Observatory Pinafore, 1879

Gilbert and Sullivan's *H. M. S. Pinafore* had premiered in 1878, and was still being performed widely in 1879 when HCO assistant astronomer Winslow Upton wrote this spoof. Its performance at HCO was canceled, probably due to the death of a member of the Observatory's staff; but it was eventually revived in 1929, updated and performed in costume by that era's astronomers and "chorus of computers." The libretto is evidence of the culture of the HCO, both when it was written and when it was performed.

26—Preparations, 1879
27—Journey Home, 1879
28—At Sea, 1879
29—Scotland, 1879

Mina did not write, or at least did not preserve, notes on her journey to Scotland to give birth. These scenes are imagined, based on the likely course of her trip.

30—Old Wives, 1879

Scotland's census indicates that Mina's mother, grandmother, and younger brother were living in the family home on Alexander Street in the city of Dundee, and renting a room to a young woman named Margret Lindsay.

The "old wives'" methods for determining the sex of a fetus are part of maternity lore.

31—Time, 1879

32—Sea of Pain, 1879

33—Birth, 6 October 1879

It is very unlikely that any young woman in 1879 would have written about her experience of childbirth, and there is no evidence that Mina did so. The scenes are imagined, but the date and place of Mina's son's birth is on record.

34—A Name, 1879

"Christian" names, first names, in the Stevens family were often repeated across or within generations. Babies were often named after admired family members (several of her siblings later named babies after Mina). It would not have been unusual for Mina to name her baby after her benefactor, Director Pickering. (Edward Fleming's name has prompted some researchers of HCO to speculate that Director Pickering might have been the actual biological father of Mina's son. There is no evidence that Mina knew Pickering before her employment there, and no evidence of scandal or innuendo in records from the Observatory, nor from any of the women working there. Mina's and Pickering's relationship seems to have been purely professional and mutually respectful, and it did not change after the deaths of either James Fleming or Lizzie Pickering. Director Pickering made no mention of Edward in his will. Admiration and appreciation appear to me, therefore, to be the reasons for Mina's choice of name for her son.)

35—The Atlantic, April 1881

Away from the Observatory, there are no records of Mina's life from the time she left HCO for Scotland and when she arrived back almost two years later, except for the Scotland 1881 census, on which she appears living in her family home in Dundee. I chose to pick up her story, then, as she returned to the U.S.

36—For the Best, 1881

Mina's thoughts here are imagined. Many immigrants at this time (and today) left family behind until they established themselves in the U.S.

37—A Women's Room, 1881

This description is based on photographs from HCO.

38—The Computers, 1881

The women's wages are part of the historical record. Director Pickering was able to hire more women because he did not pay them as much as men.

Miss Annette "Nettie" Farrar worked at HCO from 1881 to 1885. She left Cambridge, and astronomy, when she married.

39—Frugality, 1881

Mina wrote about the challenge of surviving on a woman's wages, but was noted as having a welcoming home.

40—The Work, 1881

It did not take long for Director Pickering to find that Mina was diligent and exacting in her work, and he expanded her assignments to take advantage of her abilities.

41—Missing Him, 1882

It appears that no correspondence was preserved between Mina and her mother and son from this time, so her thoughts here are imagined.

42—Missing Another, 1882

The fate of James Fleming is uncertain. Although there were several men with his name living in the area at the time, it seems likely that Mina's husband had returned to Scotland and may have passed away by this time. Mina did not write or speak about him, or at least did not preserve any record of doing so. Her thoughts here, of a general absence felt, more than a love lost, are imagined.

43—Photography, 1882

Pickering was an early proponent of using photography to aid astronomy, and he was determined that Harvard become the foremost authority on photometry. In a few years, astronomers' knowledge of the heavens expanded exponentially as the light of the stars could be recorded on photographic plates rather than observed with human eyes.

44—An American Game, 5 November 1882

45—The Score, 1882

Mina's enthusiastic support for the "Harvard Eleven" was well-known at the Observatory. Holmes's Field was used until a stadium was built; the game of football was still evolving, and it was decried as being quite violent. This game was reported in the local newspapers, and the score, unfortunately, is accurate.

46—Transit of Venus, 6 December 1882

This account is based on reports from HCO and the Boston newspapers from this time.

47—A Visitor, February 1883

Mrs. Anna Palmer Draper (1839-1914), the daughter of a real estate and railroad tycoon, had married Dr. Henry Draper, a physician, chemist, astronomer, photographer, and telescope maker, in 1867. She assisted and traveled with him for his astrophotography hobby, and was determined to see his work carried on after his death. Although the conversation that Mina overhears is imagined, many such conversations, as well as correspondence, took place before Mrs. Draper allowed Harvard to take over her husband's work.

48—Indulgences, 1883

Prices here are based on data from the time.

Andrew Carnegie, a Scottish immigrant who became the richest man in America, was a hero to many immigrants.

49—An Accounting, 1884

Gaps in funding were frequent issues at HCO; the timing of this one is noted in records from the time. Some assistants and computers often continued working without pay, hoping it would soon be restored, and it often was.

50—An Education, 1884

Pickering assigned Mina to look for errors in earlier astronomers' assumptions and calculations, which certainly led her to understand more of the history and science of astronomy, as well as the mathematics.

Ptolemy (100-170) was a mathematician and astronomer in Egypt of the Roman Empire who produced an early star catalog and maps of constellations as well as calculations for predicting astronomical phenomena. William Herschel (1738-1822) was a German-British astronomer who produced catalogs of nebulae, studied "double" stars, and discovered Uranus. Much of his work was done in collaboration with his sister, Caroline Herschel (1750-1848), who herself discovered several comets.

51—Invitation to Tea, May 1885

This specific scene is imagined, although there were many such meetings with Mrs. Draper, and the hotel was a likely place for her to stay when she visited from New York. The Parker House is still open today. At the time of this story, it had recently hosted Charles Dickens, Longfellow, and Nathaniel Hawthorne.

52—Stories in the Stars, 1885

Mina is credited with a discovery in each of the constellations included here, as well as many more.

53—Miss Farrar's Discoveries, January 1886

Miss Nettie Farrar (1862-1948) (later Mrs. Nettie Farrar Harris) studied stellar spectra for the Draper project.

The Pleiades cluster, now known to be more than 800 stars, is in the constellation Taurus. The name is from Greek myth: the seven sisters, daughters of Atlas, were so distressed when Atlas was condemned to hold up the heavens on his shoulders that they joined him in the sky.

54—Waiting, 1886

Mina's thoughts here are imagined, and the reasons for the delay in Edward
joining her are my assumptions.

55—A Romance, 1886

This particular gathering is imagined, although Mina's home was often
a gathering place for friends and family, and she enjoyed entertaining
(on a budget). Magic Lanterns were popular entertainments at the time,
and photographs on glass plates had replaced painted images as their
content. Shops advertised lanterns and slides for sale, and Mina later sent
a slideshow and a script which she wrote herself to the Carnegie family
for the amusement of Margaret Carnegie and her cousins.
Very few of the women computers during Mina's time married.

56—The Henry Draper Memorial, 1886

Anna Draper's decision to allow HCO to continue her husband's and
her work, and to finance it, was instrumental in making Harvard an
authority on astrophotography.
Mina's tired arm here is a reference to what would become a persistent
problem for which she received physical therapy. It is probably a
repetitive stress injury from her work.

57—Otherwise Engaged, Dec 1886

Director Pickering assured Mrs. Draper that Mina would be able to take
over Nettie's work with the star spectra.

58—Sorrow and Joy, 1887

Mina recounted the story of her time in the hospital, and her family
heritage, to friends.
Genealogical data matches her grandfather's birth in Spain during the war.
Since her grandmother's death preceded Edward's and his grandmother's
departure by months, it seems likely that Mina's mother was caring for
her own mother and was unable to leave Scotland until she passed away.

59—Portrait, 1887

The inspiration for this scene is an actual portrait of Mina.

60—More Work, More Workers, 1887

Wages listed here were accurate at HCO at the time. Women were hired because they proved competent at the work they were asked to do at HCO, but also because it was possible to hire more of them with the same limited resources.

Mina was at one time tasked with investigating a possible dormitory for the women computers. It was ultimately decided that women living on campus (who were still not allowed to be students there) would be distracting to the male students.

61—Form of Records for Harvard College Observatory, 1887

The italicized passages are direct quotations from the document in Harvard's files. A consistent method of recording data made the records available and readable for all, and made it possible for any computer to pick up where another's work left off.

62—The Colors of Light, 1887

Absorption lines are black lines of varying width and strength which are visible in the spectrum of a star. Gases (usually hydrogen) in the outer layers of stars absorb some of the light from the star. Because Nettie and Mina (and later others such as Annie Jump Cannon) were looking at the spectrum on photographic negative plates, or on positive black-and-white prints (color photography not coming until later) they were identifying patterns of lines that look rather like tiny versions of today's retail barcodes.

German physicist Joseph von Fraunhofer (1787-1826) first identified the lines, which came to be called **Fraunhofer lines**.

Robert Bunsen (1811-1899) and Gustav Kirchhoff (1824-1877), working in Heidelberg, used a spectroscope they invented to identify elements by their spectra. They discovered cesium and rubidium. The Bunsen burner, used in chemistry labs today, aided in their experiments.

Sir William Huggins (1824-1910), with his wife Lady Margaret Lindsay Huggins (1848-1915), pioneered the use of spectroscopy with stars.

Angelo Secchi (1818-1878), a Roman Catholic priest and physicist, directed the Observatory of the Jesuit College of Loreto, where nuns read spectrographic plates of stars. He developed the first classification of stars by their spectra.

63—Emigrants, 1887

Records of the ship S. S. Prussian list these passengers in the aft section of the ship, to become "settlers" in America: Mary Stevens, housewife, age 50 (Mina's mother); Andrew Stevens, age 11 (I assume him to actually be Andrew Davidson, son of Mina's sister Mary Stevens Davidson); Joanna Stevens, age 11 months (assumed to actually be Johanna Mackie, baby daughter of Mina's sister Johanna Creighton Stevens Mackie); and Edward Fleming, age 8 (actually still 7 years old at the time).

64—Preparations, 1887

Mina's preparations are imagined, although Mina did deal with Clark & Sons in Boston. Alvan Clark (1804-1887) and his sons George Basset Clark (1827-1891) and Alvan Graham Clark (1832-1897) became expert in constructing lenses for telescopes, and built what were the largest refracting telescopes in the world at the time.

65—My Bonnie Lad, 30 September 1887

This scene is imagined, although the date is accurate for Edward's arrival.

66—Connected, 1887
67—Echoes, 1887

These scenes are imagined. Although there are not photographs available of Edward at this age, photos from his college years show quite a resemblance to his mother Mina.

68—Classifying, 1887

As this was several years before astronomers understood the significance of the absorption lines, Mina was essentially grouping stars by the appearance of their spectra alone. She used letters rather than numbers, thinking that as new theories and discoveries might reveal some other sort of order, letters could be arranged in any way, while numbers would imply an existing order. The classes were eventually reordered, necessitating mnemonic devices for their memorization by student astronomers ever since.

69—His Eighth, 1887

I imagine Mina making up for the birthdays she has missed in this scene. Prior to any toy cars, planes, or even wax crayons, marbles seemed a likely toy. Edward did later play football for his class at M. I. T., as well as some field events for his class track team.

70—Birthday Party, 1887

The birthday party is an assumption, but the Bailey family and Mina's siblings and their families were frequent guests.

Mina and her son were members of the First Church in Cambridge (Congregational) at the corner of Garden and Mason Streets, walking distance from their home on Upland Road.

Edward's interest in bits of rock and metal here are my invention to foreshadow his interest in metallurgy.

71—Enter Miss Maury, 1888

Antonia Coetana de Paiva Pereira Maury (1866-1952) eventually devised her own classification system for stars. She was the niece of Anna and Henry Draper, and maintained a testy but collaborative relationship with HCO even when she took time away.

72—Uncle Charlie, 1888

Although there is little evidence about the personality of Mina's brother Charles Stevens (1863-1920), he gave her address as his destination at immigration, and was listed as a watchmaker. Mina spent more time with him, both in Dundee and in Cambridge, than with any other of her siblings (except for the later times at work with her sister Johanna C. S. Mackie). They seem to have been together at several trying times in both of their lives, and Charles named three of his children, in various ways, after his sister. While some of the details of their relationship are imagined throughout the story, I assume it to be a warm one like my

own between older sister and younger brother, and between uncle and
nephew (as it later was between Mina and her nephews).

73—Binary Stars, 1889

The discovery of "spectra-scopic binary" stars was a leap forward, and led
to the reexamination of previously identified stars to determine whether
they might also be binaries.

74—Another Heiress, 1889

Catherine Wolfe Bruce (1816-1900) was a New York heiress who inherited
part of the fortunes of her parents as well as that of Catharine Lorillard
Wolfe, a benefactress of The Metropolitan Museum of Art. Miss Bruce
was frequently in ill health, but developed an interest in astronomy late
in her life.

75—How I Wonder What You Are, 1889

This song was first published as a poem in 1806, in the book *Rhymes for
the Nursery* by Ann and Jane Taylor in London, and later was set to
an arrangement by Mozart of a French folk melody, also used for the
alphabet song and "Baa, Baa, Black Sheep." The song was newly popular
at this time.

The idea that stars might be masses of gas was still years away from
acceptance; most scientists assumed that stars were made of the same
stuff as Earth.

76—Southern Sky, 1889

Solon Bailey (1854-1931) of HCO established Boyden Station near the
volcano El Misti in Peru in 1890, to observe and photograph the stars
not visible in the northern hemisphere.

77—My Work, 1890

The Draper Catalogue of Stellar Spectra consists of hundreds of pages
of tables, two of which list more than 10,000 stars each, one with 9
and one with 7 columns of data assembled for each star. Mina would
have handwritten each of these entries, in addition to having previously
computed many of the values entered there, as well as writing 111 pages

of double-columned "remarks" explaining any data or observations out of the ordinary for each entry. She would have then checked and approved the printer's copy before seeing the finished publication.

78—Recognition, 1890

Mina would not have complained about the brief recognition of her work; it was customary for the college, the benefactor, and the director of an observatory to get credit for the work done there, and to be the public-facing representatives of any scientific advances.

79—A Raise, 1890

It was about this time that Marie (sometimes referred to as Mary) Hegerty, an Irish immigrant, began her employ in Mina's household.

80—On the Plates from Peru, 1890

Mina's 1890 observation of "a semicircular indentation 5 minutes in diameter 30 minutes" from Zeta Orionis (a star in Orion's belt) is the first written discovery of what later became known as the Horsehead Nebula. She was looking at a plate taken in 1888.

81—Exit Miss Maury, 1892

Antonia Maury and Director Pickering retained a mutual respect even as she felt a need at times to escape the pressures of her work and feared leaving it to others and losing credit for it.

82—Brick Building, 1892

More storage space was required for the quantity of glass photographic plates, and a building made of brick, rather than wood, would protect them from the very real danger of fire. Horse-drawn, hand-pumped fire wagons, along with early fire extinguishers, were the only defense.

83—The Transit of the Plates, 1892

This is actually how the plates were transported from one building to the other.

84—A Lesson From the Dolphin, 1893

Mina's HCO notebook records her measurements for this star on January 3, 1891. When it was contested, the two challengers each measured the star and met to corroborate their findings, only to discover that they had each been tracking different stars, and neither was the target star. Mina referred to their error in her address read at the Exposition in Chicago in 1893.

85—A World's Fair, 1893

The Columbian Exposition was organized to celebrate the 400th anniversary of Christopher Columbus's discovery of America (as it was celebrated then) in 1492 (the fair was held a little late). There was an exhibition hall specifically celebrating the accomplishments of women. Since Mina did not attend, her address was presented by physicist and astronomer Dr. William Pickering (1858-1938), Director Pickering's brother.

86—Insult, 1893

Seth Carlo Chandler (1846-1913), who had worked at HCO from 1881 to 1886, published his own variable star catalog in 1893. Resisting the trend of discoveries using spectral photography, he omitted all of Mina's recent discoveries because they had not been observed with the human eye (via telescope). He also labeled more of her discoveries "alleged but unconfirmed." Although the press stoked the controversy, Pickering and Mina acknowledged a few statistical errors but essentially ignored the insult publicly and of course were eventually vindicated.

87—"A Field for Women's Work in Astronomy," 1893

These are entirely Mina's words.

Caroline Herschel (1750-1848) was a German astronomer, the sister of astronomer William Herschel, and the discoverer of several comets.

Mary Somerville (1780-1872) was a Scottish scientist and polymath who studied the solar spectrum and magnetism, and predicted the discovery of the planet Neptune.

Maria Mitchell (1818-1889), born in Massachusetts, was the first American woman astronomer and astronomy professor, at Vassar.

88—Letters, 1893

Mina wrote many letters such as these on behalf of Director Pickering and herself. Astronomers around the world eventually wrote directly to her, as well as the Director, for information from the glass plate collection.

89—Thousands of Stars, 1893

Ruth Bailey and their son Irving lived in Peru while Solon Bailey directed the observatory there. They socialized with Mina when they were in Cambridge.

In Mina's time, **nebulae** were just coming into focus well enough, through stellar photography, that stars could be discerned in their blurriness. Today, a **nebula** is thought to be a cloud of interstellar gas and dust.

90—Nova Normae, 26 October 1893

This scene is based on the actual discovery by Mina. She is credited with the discovery of Nova Normae on this date.

91—The Bruce Telescope, 1893

Dava Sobel described the assembly of the telescope in *The Glass Universe*. Thanks to Dr. Scott Shaw, Astronomy Professor Emeritus at The University of Georgia, for describing the Bruce telescope to me.

92—Where Do the Years Go?, 1893

This is an imagined scene, although Mina did begin teaching at the age of 14, and her household in Dundee was populated as described.

Author Edgar Allan Poe (1809-1849) was born in Boston. His short stories starring the fictional character Auguste Dupin, "The Murders in the Rue Morgue" (1841), "The Mystery of Marie Roget" (1842), and "The Purloined Letter" (1844) are considered the first detective stories.

93—Harvard versus Penn, 30 November 1893

William Henry Lewis (1868-1949) already had a degree from Amherst when he played football while in Harvard Law School. He was the first Black All-American football player, and one of the first Black members of the American Bar Association. He coached football at Harvard, wrote books about football, and was eventually appointed Assistant Attorney

General in 1910 by President Taft. Lewis's profile, the description of the game, and the weather of the day were all reported in Boston newspapers.

Years later, Edward did play fullback on his class level's football team while in college at M. I. T.

94—An Ache, 1893

Pleas for Mina to take care of herself, and to take time off, are numerous in the correspondence that she preserved. She describes receiving physiotherapy for her sore shoulder in her journal of 1900.

95—Enter and Exit Miss Maury, 1894

Mina describes the preparation of other astronomers' work for publication (a painstaking and time-consuming task) as part of her duties that kept her from the work she most liked: hunting for stars on the photographic plates.

96—Misses Leavitt and Cannon, 1897

Henrietta Swan Leavitt (1868-1921) worked as a volunteer at HCO for 7 years, and eventually became a paid member of the staff in 1902. Her discovery in 1912 of the period-luminosity relationship of variable stars, now known as **Leavitt's Law**, made it possible for astronomers to calculate the distances of distant galaxies.

Annie Jump Cannon (1863-1941) also started as a volunteer, but became a paid member of the staff in 1897. She became expert at classifying stellar spectra, and eventually reordered and modified Mina's classification system in a way that is still used today.

97—An Educated Man, 1897

Mina worried about the expense of Edward's education in her Journal of 1900. There is no evidence that she received any assistance in paying it from her own salary.

Edward's initials appear in the computers' notebooks at HCO; he probably worked for hours at a time on breaks from school.

Boston and Cambridge were served by horse-drawn trolleys and cable cars in the mid-1800s, and the first subway in America opened the Tremont Street Station in 1897.

At this time, Mina's sister Johanna Creighton Stevens Mackie and her husband James Mackie, a tailor, lived in Boston; her brother John, an instrument-maker and engineer, and Charles, a watch-maker, lived in Cambridge.

98—A Request, 1897

Solon Bailey reported about his attempts to protect the telescopes through earthquakes and civil war. Correspondence between Bailey and Pickering described the complaints of Mina, on behalf of the computers, and the assistant astronomers, who were preparing plates, maintaining and setting the telescopes, and taking the photographs in Peru. The compromise they reached was to acknowledge the assistants for their work in photography when the announcements of discoveries were made in Harvard publications.

99—The Amusing Story, 1898

The story that Pickering, exasperated with his male assistants, exclaimed, "My Scottish maid could do better!" has persisted in the tale of the Harvard Women Computers. That the story actually came later, after Mina had achieved success, is my theory.

100-the Bruce Medal, 1898

The Catherine Wolfe Bruce Gold Medal has been awarded almost yearly since 1898 "for outstanding research in astronomy." Director Pickering received the award in 1908. Simon Newcomb (1835-1909), the first recipient of the award, was a Canadian-American self-taught professor of Mathematics and Astronomy. He studied the speed of light, latitude, and a variety of other scientific studies, and his recalculation of astronomical constants became the standard in international astronomy.

101—An Astronomical and Astro-physical Society, 1898

George Ellery Hale (1868-1938), an American solar astronomer specializing in sunspots, proposed that Pickering host this meeting at Harvard. They and Simon Newcomb realized a need for a professional organization of astronomers, and a debate was arising about whether astrophysics should be included. The Astronomical and Astrophysical Society of America, now the American Astronomical Society, was not established until the following year, in 1899.

Arthur Searle (1837-1920) began as a computer at HCO, became an assistant astronomer, and served as interim director for 2 years before Pickering was hired as Director. He also wrote a history of the HCO.

An account of this meeting, including the heat wave, Mrs. Pickering's hosting, and Mina's enthusiastic reception, was published contemporaneously in *Popular Astronomy*.

"Astro-physics" seems to have appeared earlier, and being a newly-coined word, it eventually gave way to "Astrophysics." Both were in use at this time.

102—A Title, 1899

In 1899, Mina became the first woman to hold a title at Harvard, Curator of Astronomical Photographs. She now was an employee of Harvard University, not just the Observatory.

103—Photo Gallery

104—The Turn of the Century, 1899-1900

This scene is imagined. The predictions for the next century were still beyond the horizon at the time.

105—A Journal, March 1900

Mina's handwritten journal is part of Harvard's archives, and can be read online. It is the major inspiration for her unique voice for this work.

106—"Journal of Williamina Paton Fleming / Curator of Astronomical Photographs, Harvard College Observatory," March 1900

The italicized introduction is in Mina's own words. The rest are summaries and convenient arrangements.

107—The Journal: Evenings Out, March 1900

This is a collection of Mina's accounts of entertainments and visits with friends and family during the first few weeks of her journal-keeping.

The elegant and ornate Castle Square Theatre was on Tremont Street in Boston until 1932.

Arthur Conan Doyle's most famous character, Sherlock Holmes, had not yet finished his investigative career at this point. *Girdlestone* was based on one of his non-Holmesian novels.

Annie Cannon did board with Mina for several years; the house was within walking distance of the Observatory, where Cannon worked at the telescope many evenings.

Some of Mina's friends were prominent in Boston society. Mrs. Atherton Brown of Roxbury was a patroness of the arts. Captain Adams of the Cable Ship *Minia* lived in Boston when not at sea.

Mina was known to keep small gifts for children in a drawer in her office, ready to be given to young visitors or children of her colleagues.

108—The Journal: Evenings In, March 1900

This is a collection of the evenings' activities that Mina described in her Journal of 1900.

The quotation marks around "news" in referring to reading *The Herald* are Mina's, and may reveal a skepticism or frustration with dealing with the press.

The "Cuban women" were a charity for which she made and sold items, perhaps a church concern.

She sent dolls (some designed to embody the characters of particular stars) and toys to the families working in Peru for their children and local youngsters for Christmas celebrations.

She mentions "India," a rummy-type card game, several times in her papers; hot cocoa and fudge appear as well. Her "spree" of treats seems to be a buffet. There are recipes for creamed walnuts from the time. She set up games of jackstraws (pick-up sticks), crokinole (a type of tabletop carom board), and cue ring (a sort of ring-toss) for her guests to take turns at.

Alone some other nights, she described herself as "the lodestone left to keep
the house from blowing away."

109—The Journal: Reflections on a Fortnight of Work, March 1900

The italicized notes are Mina's words, rearranged here. In her Journal they
appear as a day-by-day narrative until they taper off when she becomes
ill. Some tasks appear multiple times over the first 2-week period of the
Journal,

Her frustration is apparent: "If one could only go on and on with the
original work, life would be a most beautiful dream . . ." However,
she was bound by the work Pickering assigned to her: "you have to
put all that is most interesting to you aside, in order to use most
of your available time preparing the work of others for publication."
Although she was finding new variable stars and nebulae, her value to
Pickering was administrative, secretarial, and organizational. She put on
a brave and determined face, however (although she may have already
been feeling ill, and getting more frustrated and overwhelmed by her
assignments as "the Grippe" took hold). She quotes from the Bible's
Ecclesiastes: "However, 'whatsoever thy puttest thy hand to, do it well.'"

110—The Journal: A Tour, March 1900

Mina seems, from the words in her journal, to have been both proud that
her son wished to bring visitors to HCO and a bit put off that he did not
give her much notice. It's probable that she was also feeling overworked
and possibly ill at this time.

Edward Skinner King (1861-1931) was in charge of photography at HCO.
He later became a professor of astronomy at Harvard University.

111—The Journal: A Personal Reflection on the Value of a Woman's Work, 1900

These are exactly Mina's words, speaking up for herself.

112—The Journal: In Which the Activities of Work and Home Grind to an Unpleasant Halt, March 1900

Aside from the title, these again are all Mina's words. Her exhaustion and
illness are apparent.

The influenza virus had not been identified in 1900; the Grippe was the common term for its symptoms (fatigue, muscle aches, headache, sore throat, cough). Bed rest, tea, and broth were the primary treatments; the grippe sometimes worsened and patients developed pneumonia.

113—The End of the Journal, and Very Nearly of Me, March 1900
Recovering, Mina looks back at her journal. The italicized words are her own, and the end of her journal assignment for Harvard.

114—A Last Reflection on the Journal Accompanied by an Apology of Sorts, 18 April 1900
Entirely Mina's words, this note, added as she handed over her journal for posterity, again captures her voice and dogged determination to make things better for herself, her family, and other working women.

115—A Nomination, 1900
Catherine Wolfe Bruce died on March 13, 1900. Mina did not win the Bruce Gold Medal that year.

116—A "Field Trip" to Georgia, May 1900
A **solar eclipse** occurs when the Earth, Moon, and Sun are aligned such that the Moon blocks the light of the Sun from parts of the Earth. Washington, Georgia, unlike Cambridge, was in the **path of totality**, the area for the 1900 eclipse in which the Sun would be nearly completely blocked by the moon.

117—An Adventure, May 1900
Mina already knew Mr. Adams, the *Minia*'s ships officer. This cable ship laid and repaired trans-oceanic telegraph lines. The dog, who lived aboard the ship for 9 years, was named Rover. The *Minia* was later one of the ships that recovered bodies from the *Titanic* disaster.

118—In the Field, May 1900
Wilkes County, Georgia was a center for cotton production in Georgia where, less than 40 years earlier, enslaved people had worked the cotton

fields. The city of Washington had a population of more than 3000, and
a stately new Gilded Age hotel had recently opened on its square.

In addition to Edward Fleming and Mrs. Draper, these women computers
were on the trip with Mina: Miss Evelyn Leland, Miss Louisa Wells, Miss
Florence Cushman, Mrs. Imogene Eddy, Miss Ida Woods, Miss Edith
Gill, Miss Mabel Gill (sisters), and Miss Mabel Stevens. It is possible
that others were there, possibly assisting William Pickering who arrived
earlier to set up camp.

119—Early Morning, Washington, Georgia, 28 May 1900
120—Totality, 28 May 1900
121—Shadows, 28 May 1900
122—Final Contact, 29 May 1900

These scenes are based on newspaper accounts from reporters on site, as
well as my personal observations of solar eclipses and the landscape of
this area of Georgia.

123—Back to Work, June 1900

Although astronomers were calling it Witt's Planet after its discoverer,
Eros was not a planet, but an asteroid. Seth Carlo Chandler worked out
the orbit of Eros and narrowed down dates and locations where it had
passed Earth in earlier years. Although he had ignored Mina's discoveries
years earlier, he now requested that she examine the plate collection to
see if some trace of the asteroid might have been visible then. Mina was
able to locate it on plates from 1893, 1894, and 1896.

Director Pickering and other astronomers were especially interested in
Eros because it would be passing close enough to Earth for them to use
its position to calculate distances in space, including the sun's distance
from the Earth.

Again, Pickering asks Mina to put aside her examination of the plates and
concentrate on shaping for publication the data the astronomers were
gathering in increasing amounts and at increasing rates.

124—Lady Assistants, Autumn 1900

Although the credit Mina was given for the work done at HCO may seem meager, it was steadily (but slowly) increasing, as was the acknowledgement of the other women's work.

125—Miss Cannon's Classification, 1901

Cannon's OBAFGKM order is still memorized by astronomy students today, who may use the mnemonic later popularized by Henry Norris Russell's Princeton students: "Oh, Be A Fine Girl, Kiss Me." Mina would certainly have had a thing or two to say about that.

126—A Graduate, June 1901

This is an imagined scene, but a celebration was certainly in order.

"Slàinte mhath" is a traditional toast in Scotland and Ireland (where it's spelled a little differently), from Gaelic. It is pronounced something like "slanj-a-va" and translates to "good health."

127—Another Nomination, 1901

Another nomination for the Bruce Medal, again without success.

128—The Grippe, 1901

Mina suffered influenza again at about this time.

129—An Anniversary, 1 February 1902

This celebration did actually occur, and the gifts to the Director were as described here.

130—Dinner with Astronomers, 1902

The American Astronomical and Astrophysical Society (now the American Astronomical Society) was more inclusive than many professional organizations at the time. By 1910, its admitting class was one-sixth women.

Newcomb's words are his own.

131—Building, 1902

After all of Mina's fears of fire, it was actually a flood in later years (after
  Mina's time) that threatened to destroy the collection of astronomical
  plates.

132—A New Benefactor, 1903

Andrew Carnegie (1834-1919) had immigrated from Scotland at age 12.
  He made his fortune in railroads and steel, and followed his own "Gospel
  of Wealth" which called on the wealthy to use their money for the good
  of society.

133—A Provisional Catalogue, 1903

Mina had written in her Journal of 1900 of her continued prompting of
  Miss Cannon to finish the remarks for this catalogue, which Mina was
  preparing for the printer.

134—The Return of Miss Leavitt, 1903
135—Invention, 1903

Some of Miss Leavitt's fly spankers are in Harvard's collection.

Wolf-type variables, now **Wolf-Rayet stars**, rare groupings of stars many
  times more massive than the sun, with unique spectra, burn extremely
  hot and bright.

136—An American Citizen, 1904

Edward became a naturalized citizen on January 12, 1904.

137—Room for More, November 1904

Charles Stevens's first wife, Eliza(beth) Margaret Kerr Stevens, died
  November 12, 1904. They had married in 1893.

138—Losses, 1905

It is not known when Mina actually learned of her husband's death, but
  she and Edward had described him as "deceased" on records for several
  years.

Her sister Mary was the only one of Mina's siblings who remained in
  Scotland. Several of Mary's children came to America and settled with
  their aunts and uncles in Cambridge and Boston.

Mina's early miscarriage seems not to have been common knowledge, but she suggests it in correspondence with Louise Carnegie and she numbered 2 pregnancies and 1 live child on a census.

139—Nebula, 13 January 1905

This part of a nebula is now known as "Fleming's Triangular Wisp." It was originally called by Pickering's name.

140—Once Again: Cambridge, 1905

Another nomination for the Bruce medal. Pickering's words (italicized) come from his nomination of Mina.

141—At Wellesley, 1905

Sarah Whiting (1847-1927) was a professor of physics and director of the observatory at Wellesley. Mina was a frequent guest lecturer, but this particular scene and the questioning student are imagined.

142—Corresponding with Mrs. Carnegie, 1905

The conversation with Director Pickering is imagined, but Mina preserved some of her correspondence with the Carnegies.

143—Star Library, 1905

There are more than 2500 Carnegie Libraries throughout America and in other countries of the world.

The Sibylline Books were Greek texts allegedly from the oracles of ancient Greece, consulted by Roman emperors in times of crisis.

144—A Reply, 1906

Mina preserved at least some of her letters from the Carnegies, as well as drafts of hers to them.

145—A Sky Map, 1906

While Director Pickering welcomed the discovery of new stars, he was particularly concerned with making HCO the repository for astronomical plates and data that would verify and stimulate astronomical discoveries all over the world.

146—College Women, 1906

Elizabeth Cary Agassiz (1822-1907), wife of a Harvard professor, originally educated young women in her home, then began a facility for women affiliated with Harvard and known as the Annex. Radcliffe was a full-fledged college for women, which is now the Radcliffe Institute for Advanced Study at Harvard University.

147—Valentines, 1906

Mina described making Valentines with her nephews in a letter to Louise Carnegie. Charles's request of a Valentine for Miss Honan is imagined, foreshadowing his romantic interest in her.

148—An Honor, May 1906

Annie Jump Cannon had a significant hearing loss, probably as a result from the scarlet fever she suffered as a young woman.

The Director designed and had built a large, multi-sided rotating desk, among other useful innovations.

Mina was the sixth woman, and the first woman working in America, to become an honorary member of the Royal Astronomical Society. They were "honorary" because women at the time were not allowed to be full members.

149—Mrs. Pickering, August 1906

Lizzie Pickering died August 29, 1906. She was 57 years old. Mount Auburn Cemetery is the final resting place of many prominent Cambridge and Boston citizens, including Edward Pickering and, eventually, Mina.

150—A Fellow, 1906

Mina described this incident to a Boston reporter.

151—Publicity, 1906

These scenes are based on the interview reported in the Boston Globe. The italicized words are directly quoted from that article.

152—Headlines, 1906

Many newspapers ran the same story, or an edited version, with their own headlines. Each of the headlines and newspaper names is authentic.

153—Miss Cannon's Catalogue, 1907

Cannon was apparently little hampered by her hearing loss. She also reportedly did not mind the physical work of climbing ladders, pulling ropes, and turning gears to manipulate the telescopes.

154—A Marriage, 1907

Widower Charles Stevens and Josephine Mary Honan (1879-1940) were married April 23, 1907. They had at least 6 children together, as well as 2 sons from Charles's first marriage. Miss Honan had been taking care of Mina's and Charles's mother, Mary Walker Stevens.

155—Fire, 4 March 1907

Mina described her fire-fighting adventure in the draft of a letter to Margaret Carnegie, Louise and Andrew Carnegie's young daughter.

156—Marvelous Work, 1907

Herbert Hall Turner (1861-1930) was the director of the Radcliffe Observatory at Oxford University. The italicized words are his.

157—American Woman, September 1907

Mina became a naturalized citizen on September 9, 1907.

158—A Letter from Mr. Carnegie, 1907

The letters referenced here are in Mina's papers at Harvard.

159—Medals, 1908

Pickering was the seventh recipient of this medal, although he had himself nominated Mina. "I cannot do better than repeat my recommendation" of her, he had written in one of his nominations.

160—Rest, 1909

Mina Davidson was the daughter of Mina Fleming's sister Mary Stevens
Davidson. She had immigrated as a child with some of her siblings in
1888.

Louise Carnegie's words are her own in a letter saved by Mina. Mina's
illness was recounted to her in another letter.

161—The Works of Misses Maury and Leavitt, 1909

These women both made important contributions to astronomy, and
eventually became better known and respected in the field than Mina
had been.

162—The Return of Mr. Halley's Comet, 1909

Maximilian Wolf (1863-1932) was a German astronomer and
astrophotographer and the discoverer of numerous stars and asteroids,
including Brucia, named for Catherine Wolfe Bruce. (It is Charles Wolf,
not Maximilian, for whom Wolf-Rayet stars are named.)

Sherburne Wesley Burnham (1838-1921) was a professor of astronomy at
the Yerkes Observatory of the University of Chicago.

163—Group Photograph at the Yerkes Observatory, 1909

This scene is based on the group photo from this conference. Although the
University is in Chicago, the Yerkes Observatory was actually nearby on
Lake Geneva in Wisconsin.

164—Farther West, August 1909

This specific scene is imagined, but Mina did make this trip, and Edward
apparently alerted the local newspaper that his famous mother was
visiting; it was reported there.

165—Three Girls from New York, December 1909

Mina kept Mrs. Costikyan's thank-you letter for hosting her daughters, in
which she mentions the rug that she is sending. The Costikyan family's
rugs were in some of the finest homes and businesses in New York and
Boston.

166—Another Decade, January 1910

Mina's thoughts are imagined here, but they are based on an entry in one
of her HCO notebooks from this time.

167—Collecting Stars, 1910

Mina's thoughts here are imagined.

168—The Comet, May 1910

This is based on contemporaneous reports from the *Boston Globe*.

169—The Puzzle of Omicron² Eridani B, August 1910

Henry Norris Russell (1877-1957), a Princeton astronomy professor,
described this scene, the discovery of what came to be called a white
dwarf, including the note to Mina and the words of Pickering. A **white
dwarf** is a star that has spent its nuclear fuel, leaving a dense, hot core.

170—"The Best-laid Schemes," August 1910

The title quotes is from a Robert Burns poem, "To a Mouse": "The best
laid schemes o' mice an' men / Gang aft a-gley."

Johanna Creighton Stevens Mackie, Mina's older sister, worked at HCO
from 1903 until at least 1919. Mina's niece, Ida May Stevens, worked
there in 1904 and again from 1907-1909. She was the daughter of Mina's
older brother Robert.

The itinerary for what was essentially a joint traveling conference of
the Astronomical and Astrophysical Society and the Solar Union was
reported in newspapers and by various members.

171—The Trip Begins in Cambridge, 19 August 1910

Blue Hill Meteorological Observatory is 10 miles north of Boston, and it
holds the oldest continuous record of weather in the United States.

Sarah Whiting, the physics professor, directed the Whitin Observatory at
Wellesley.

There was a student astronomy lab down the hill from HCO, on Harvard
Yard.

172—Niagara, August 1910

Pickering kept a record of this trip, and the trip to the Falls is based on his account there.

173—Chicago, August 1910
The group did visit the University of Chicago and the Yerkes Observatory. The references to the 1893 Exposition are documented. Mina's thoughts, though, are imagined.

174—Night Train, August 1910
Mina's thoughts are imagined, but the itinerary is factual.

175—On to Arizona, August 1910
The heat was reported by Pickering and others on the trip.

176—Flagstaff, August 1910
Percival Lowell (1855-1916) was a Harvard mathematics graduate and business man who founded his own observatory in what was still the Territory of Arizona. He was especially interested in studying planets, and led the search for a planet he theorized must lie beyond Neptune, later identified as Pluto.

177—The View, August 1910
178—Farther, and Farther, West, August 1910
179—Pasadena, August 1910
180—Up the Mountain, August 1910
These four scenes are based on contemporaneous accounts of the trip and photographs taken there.

181—Mt. Wilson, August 1910
Hale's **spectroheliograph** could photograph the Sun in the light of a single element. It was still under construction in 1910.
Jacobus Kapteyn (1841-1922) was a Dutch astronomer who studied the motion of the stars.

182—The Case for Harvard, August 1910

Hermann Karl Vogel (1842-1907) had developed a system of star classification based on Secchi's earlier system. He was also a discoverer of binary stars.

Pickering appears to have been quite the diplomat, bringing astronomers and physicists of differing opinions together and guiding them toward the systems he thought would best serve astronomy (and, coincidentally, HCO).

183—Descent, September 1910

A photo of Mina with a few colleagues during a break on the trail shows her looking rather uncomfortable in dark, restrictive, conservative dress, an elaborate and heavy hat, and shoes decidedly not appropriate for the trail.

184—Banquet, September 1910

Lady Huggins was recently widowed at this time. Andrew Carnegie was a major benefactor of the Mt. Wilson Observatory.

185—Dinner with Edward, September 1910

This is an imagined scene, although Mina did visit her son, and Edward did go to Chile to work soon after it.

186—Telegram, September 1910

Mina's mother, Edward's grandmother, Mary Walker Stevens, died on September 4, 1910, while Mina was visiting Edward in Salt Lake City.

187—Taking Leave, September 1910
188—Remembrance, September 1910
189—A Generation, September 1910

These three scenes are imagined, but Mina did cut short her visit and return to her siblings in Cambridge.

190—Back to Work, September 1910

The surveys coming in to HCO were favorable toward the Draper system used by Mina and Miss Cannon in the catalogues they published, but it was still several years and more discoveries later before a single system was more widely accepted.

191—Discoveries, October 1910

These are the words of the notices posted in Harvard's Astronomical Bulletin. Bulletins had been implemented in order to publish new discoveries much more quickly than they could be published in an astronomical journal or catalogue.

192—Headlines, 1910

These are actual newspaper headlines, most from the autumn of 1910.

193—A Surprise Party, December 1910

Joel Hastings Metcalf (1866-1925) was a graduate of Harvard Divinity School, a local astronomer and Unitarian minister. He was a friend of Mina's, and later officiated at her funeral.

The party, while well-deserved, is imagined.

194—Working with Miss Cannon, January 1911

"Draper," "Harvard," and "Pickering" were all used as names for the classification system. The conversation between the two women is imagined.

195—To the Printer, February 1911

Although she had completed the preparations, Mina's final catalogue was not published until after her death.

196—The Work Goes On, March 1911

Thomas Hinsley Astbury (1858-1922) was a school headmaster by day and amateur astronomer by night.

Mina continued her careful notes in HBO notebooks through the spring of 1911.

197—Heeding Advice, March 1911

Charles Glidden planned a Boston-New York transit line via balloon. Mina had talked with him and had a trip planned for the summer of 1911.

Although she had mentioned returning to Scotland for a visit, there were
not concrete plans for the trip.
Louise Carnegie, Anna Draper, and the other computers had often
implored Mina to take a rest for her health.

198—Thirty Years, 9 May 1911

Although her last day at work is imagined here, Mina did enter the hospital
on this day.

199—A Difficult Time, May 1911

Mina had likely suffered from influenza for a while before entering the
hospital; she may have already had pneumonia by this time, as well.
Because this was before the development of antibiotics, Mina's doctors
could do little other than try to relieve her symptoms.

200—In Dreams, May 1911
201—Into the Ether, 21 May 1911

These scenes are, of course, imagined.

Acknowledgements

I had never heard of Williamina Fleming before I encountered her name while researching women scientists for a potential book project. I was immediately impressed by the sheer number of her discoveries. Here, I thought, was an underappreciated pioneer. But when I, a total novice in astronomy, learned about the Harvard Women Computers, I realized that Fleming had made her discoveries while sitting at a desk, not while gazing romantically at the stars. How can there be a story in that? Fortunately, I soon found Fleming's 1900 Journal and became convinced that I could make a story of her life. Obeying a request made of the Harvard Community to keep a month-long journal as a sort of turn-of-the-century time capsule, Fleming kept the record of her work days at the Observatory and her nights and Sundays at home. In the 20 pages of this journal, on lined paper in her own lovely handwriting, Fleming writes in great detail of her work and leisure time, and also, while coming down with a bad case of influenza, complains of the exhaustion brought on by the co-demands of her supervisory duties at work and at home, where she is raising a son, funding his college education, and keeping a home as a single mother. She also complains about her salary at Harvard in comparison to the men's. Then, in a note at the end of the journal cut short by her illness, she regrets speaking so harshly and apologizes, only to repeat the argument that she should be better compensated, and appreciated, after all. Here, I thought, was a working woman's voice that I understood, a voice that is as relevant today as it was in 1900. In researching Fleming, I soon encountered Dava Sobel's entertaining and informative book *The Glass Universe*, and discovered the entire community of women who not only supported the male scientists who were advancing the field through astrophotography, but actively contributed their own discoveries.

I tried to remain faithful to her voice as I wrote Fleming's story, putting her life and work in the context of her time. I consulted her Journal and her letters to find her voice and to invent a way for her to speak, including some Scots phrasing to convey what Annie Jump Cannon called her "delightful Scottish accent," particularly when Fleming was newly-arrived in America or surrounded by family. When I have directly quoted Fleming's or others' writing or reported speech in this text, I have italicized it. While I based the events in the narrative on actual events of her life, I had to imagine many of the thoughts, motivations, and details. For example, while her separation from her husband was reported in her lifetime, I could only imagine the circumstances surrounding it. I built her relationship with her brother Charles from the draft of a letter in which she describes making Valentines with his young sons, but I could only imagine their conversations (I have younger brothers, though, so I could make some good guesses). Her dealings with prominent astronomers, computer coworkers, and society members are based on fact, although I could only estimate her feelings about them. While a very few minor characters in this story were invented, the majority, in fact every character with both a first and last name in this text, are historical figures and are represented here as accurately as I could imagine, given my research into the people, places, and times of the story.

As to the science, writing Fleming's story has been a crash course in Astronomy and the Harvard College Observatory. I was able to see the Harvard College Observatory logbooks through Project PHaEDRA (Preserving Harvard's Early Data and Research in Astronomy) at the former Wolbach Library of the Harvard and Smithsonian Center for Astrophysics, and the glass plate negatives being digitized through Digital Access to a Sky Century @ Harvard (DASCH). I consulted Fleming's own publications in scientific journals, and many articles related to the Computers, some from their own time as well as more recent research into them and their work. I am grateful to Lindsay S. Zrull, who was then Curator of Astronomical Photographs at the Harvard College Observatory, and Maria McEachern, then Reference and Resource Sharing Librarian at the Harvard and Smithsonian Center for Astrophysics, for guiding me toward the history of the Harvard Women Computers; and to Karisa Zdanky, Astronomy Program Manager and Astronomer at the

Tellus Science Museum, and the late Dr. Scott Shaw, Professor Emeritus of Physics and Astronomy at the University of Georgia, for reading the manuscript, correcting my scientific errors, and suggesting clarifications for readers. Any factual errors in this story are unintended and entirely my own.

I am grateful for the support of my writers' group, who did not expect the astronomy lessons but still offered great suggestions and kept me on track: Debra Harden, Gail Karwoski, Gwen O'Looney, Muriel Pritchett, and Donny Seagraves. I also thank my husband Felix Vizurraga for his reading and enthusiasm for the manuscript, and for his interest in the science and the structure of this story, along with a host of other things. I dedicate this story to my mother, Nancy Dean Pluckhahn, who found her own place among the stars as I worked on this manuscript.

Bibliography

"1861 Scotland Census. Reels 1-150. General Register Office for Scotland, Edinburgh, Scotland." 1861. *Ancestry.com.* Web. 2021.<https://www.ancestry.com/discoveryui-content/view/3030419:1080?ssrc=pt&tid=176393814&pid=222291669062>.

"1880 United States Federal Census for Robert N. Stevens." n.d. *Ancestry.com.* Web. 2021.<https://www.ancestry.com/imageviewer/collections/6742/images/4241771-00264?pId=15361362>.

"1900 United States Federal Census for Charles Stevenes." n.d. *Ancestry.com.* Web. 2021.<https://www.ancestry.com/imageviewer/collections/7602/images/4113830_00794?pId=79188963>.

"1900 United States Federal Census for Johanna C Mackie." n.d. *Ancestry.com.* Web. 2021.<https://www.ancestry.com/imageviewer/collections/7602/images/4114450_00127?pId=24014641>.

"1900 United States Federal Census for John M Stevenes." n.d. *Ancestry.com.* Web.2021.<https://www.ancestry.com/imageviewer/collections/7602/images/4113830_00794?pId=79188959>.

"1900 United States Federal Census for Robert Stevens." n.d. Web. 2021.<https://www.ancestry.com/imageviewer/collections/7602/images/4113842_00807?pId=6156821>.

"1900 United States Federal Census for Wilhemina Fleming." n.d. *Ancestry.com.* Web. 2021.<https://www.ancestry.com/imageviewer/collections/7602/images/4113842_00913?pId=6162050>.

"1910 United States Federal Census for Chedia J Stevens." n.d. *Ancestry.com.*   Web. 2021.<https://www.ancestry.com/imageviewer/collections/7884/images/31111_4330077-00293?pId=109609304>.

"1910 United States Federal Census for Johanna Mackie." n.d. *Ancestry.com.* Web.2021.<https://www.ancestry.com/imageviewer/collections/7884/images/31111_4330853-00498?pId=167700997>.

"1910 United States Federal Census for Robert H Stevens." n.d. *Ancestry.com.* Web. 2021.<https://www.ancestry.com/imageviewer/collections/7884/images/31111_4330077-01220?pId=109650742>.

"1910 United States Federal Census for Williamine Fleming." n.d. *Ancestry.com.* Web. 2021.<https://www.ancestry.com/imageviewer/collections/7884/images/31111_4330077-00776?pId=11081034>.

"1920 United States Federal Census for Chas J Stevens." n.d. *Ancestry.com.* Web.2022.<https://www.ancestry.com/imageviewer/collections/6061/images/4301094_00576?pId=38978811>.

"A Distinguished Dundee Lady." *The Flaming Sword* Jan 1893: 363-4. Web.2021.<https://www.google.com/books/edition/The_Flaming_Sword/7qEcAQAAMAAJ>.

"A Famous Astronomer." *The Kenosha Evening News* 26 Sep 1906: 7. Web. 2021.<https://www.newspapers.com/image/595250666/>.

"A Famous Astronomer." *Champaign Daily Gazette* 6 Oct 1906. Web. 2021.<https://www.newspapers.com/image/668191150/>.

"A Famous Astronomer." *The Miami News* 16 Nov 1906: 11. Web. 2021.<https://www.newspapers.com/image/297339804/>.

"A Famous Astronomer." *The Windsor Star* 22 Oct 1906: 5. Web. 2021.<https://www.newspapers.com/image/500460286/>.

"A Famous Astronomer." *The Richmond Item* 13 Oct 1906: 3. Web. 2021.<https://www.newspapers.com/image/248284968/>.

"A Famous Astronomer. Mrs. Williamina Paton Fleming is a Brilliant Scientist." *Kenosha News* 26 Sep 1906: 7. Web. 2021.<https://www.newspapers.com/image/78845100/>.

Albright, Evan J. "William Henry Lewis: Brief Life of a football pioneer: 1868-1949." *Harvard Magazine* (2005). Web. 2021.<https://www.harvardmagazine.com/2005/11/william-henry-lewis-html>.

Aldama, Mariana Espinosa. "La Propagacion de la cultura cientifica de la Sociedad Astronomica de Mexico (1910-1916). Masters Thesis." November 2010. *Researchgate.net.* Web. 2021.<https://www.researchgate.net/profile/Mariana-Espinosa-Aldama>.

"All Creditors of Robert Stevens (Legal Notice)." *The Courier and Argus*(1859): 2. web. 2021. <https://www.newspapers.com/image/411014322/>.

"American Astronomical Society." Vol. VI: Yerkes Observatory. apf6-00107. Comp. Special Collections Research Center University of Chicago Library. Williams Bay, August 1909. Web. 2021.<http://photoarchive.lib.uchicago.edu/db.xqy?one=apf6-00107.xml>.

American Astronomical Society. *Astronomy Terms.* n.d. American Astronomical Society. Web. <https://skyandtelescope.org/astronomy-terms/>.

Ancestry.com. *1861 Scotland Census.* Ancestry.com Operations Inc. Provo, UT, USA, 2006. Web.2021.<https://www.ancestry.com/family-tree/person/tree/176393814/person/222291669062/facts>.

—. *1871 Scotland Census.* Ancestry.com Operations Inc. Provo, UT, USA, 2007. Web.2021.<https://www.ancestry.com/family-tree/person/tree/176393814/person/222291669062/facts>.

—. *1881 Scotland Census.* Provo, UT, USA: Ancestry.com Operations Inc, 2007. Web.2021.<https://www.ancestry.com/family-tree/person/tree/176393814/person/222291669062/facts>.

—. *Scotland, Select Births and Baptisms, 1564-1950.* Ancestry.com Operations, Inc. Provo, UT, USA, 2014. Web. 2021.<https://www.ancestry.com/family-tree/person/tree/176393814/person/222291669062/facts>.

Associated Press."Another New Star Is Added to Woman's List." *Pittsburgh Daily Post*. 15 Oct 1910: 3. Web. 2021. <https://www.newspapers.com/image/87694002/>.

Astronomical Society of the Pacific. *Catherine Wolfe Bruce Gold Medal*. n.d. Astronomical Society of the Pacific. Web. 2022.<astrosociety.org/who-we-are/awards/catherine-wolfe-bruce-gold-medal.html>.

Bailey, Solon I."Joel Hastings Metcalf." *Popular Astronomy* XXXIII.8 (1925):493-4. Web. 2021. <https://adsabs.harvard.edu/full/1925PA.....33..493B>.

Bailey, Solon. "The history and work of Harvard observatory, 1839 to 1927; an outline of the origin, development, and researches of the Astronomical observatory of Harvard college together with brief biographies of its leading members." *New York, London, Pub. for the Observatory by the McGraw-Hill book company, inc.,1931* (1931). Web. 2021.<https://articles.adsabs.harvard.edu/full/1931HarMo...4....1B>.

Baker, Daniel W. "History of the Harvard College Observatory During the Period 1840-1890." *Reprinted from The Boston Evening Traveler* (1890). Web.2021. <https://www.gutenberg.org/files/59633/59633-h/59633-h.htm>.

Berglund, Jennifer. *Historic Challenges for Harvard Women of Science with Sara Schechner, HMSC Connects! Podcast Episode 14 Transcript*. n.d. W e b . 2021.<https://hmsc.harvard.edu/historic-challenges-harvard-women-science>.

"Births (Mrs. Robert Stevens)." *The Courier and Argus* (1857): 2. Web. 2021.<https://www.newspapers.com/image/401553446/>.

Bonnell, Jerry and Robert Nemiroff. *Astronomy Picture of the Day: Fleming's Triangular Wisp*.27 July 2021. NASA.gov. Web. 2021.<https://apod.nasa.gov/apod/ap210727.html>.

Boyd, Lyle G."Mrs. Henry Draper and the Harvard College Observatory, 1883-1887." *Harvard Library Bulletin XVII (1)* (1969): 70-97. Web. 2021.<https://nrs.harvard.edu/URN-3:HUL.INSTREPOS:3736375 0>.

Cannon, Annie Jump. "Williamina Paton Fleming." *Astrophysical Journal* Nov 1911: 314. Web. 2021.<https://articles.adsabs.harvard.edu/full/1911ApJ....34..314C>.

"Center Rush Lewis." *The Boston Daily Globe* 12 Dec 1893: 6. Web. 2021.<https://www.newspapers.com/image/430702792/>.

Charles James Stevens Facts. n.d. Web. 2021.<https://www.ancestry.com/family-tree/person/tree/176393814/person/222296869168/facts>.

"Clever Women Who Study the Stars." *Marion Daily Star* 11 August 1906: 7. Web.2021. <https://www.newspapers.com/image/299462121/>.

"College Football Games. Harvard Defeated by Yale." *New-York Tribune (New York, New York) · 26 Nov 1882.* 26 Nov 1882: 2. Web. 2021.<https://www.newspapers.com/image/79207190/>.

"Comet Facts for Today." *The Boston Globe* 19 May 1910: 5. Web.<https://www.newspapers.com/clip/87817373/>.

Commonwealth of Massachusetts. "Massachusetts, U.S., State and Federal Naturalization Records, 1798-1950 ." n.d. *Ancestry.com.* Web. 2021.<https://www.ancestry.com/imageviewer/collections/2361/images/007774867_00338?treeid=&personid=&hintid=&usePUB=true&usePUBJs=true&_ga=2.61542757.875148363.1648604643-66737215.1642896561&pId=1826828>.

Commonwealth of Massachusetts, City of Cambridge. "Massachusetts Vital Records, 1840–1911.New England Historic Genealogical Society, Boston, Massachusetts. Massachusetts Vital Records, 1911–1915. New England Historic Genealogical Society, Boston, Massachusetts./p."n.d. *Ancestry.com.* Web. 2021.<https://www.ancestry.com/imageviewer/collections/2101/images/41262_b131967-00485?pId=8519454>.

Cortner, David. *Fleming's Triangular Wisp, The Starry Night, 185.* 12 November 2016. Web. 2021.<http://www.davidcortner.com/slowblog/20161113.php>.

"Death Notice LA Times - Edward Pickering Fleming...April 1962." April 1962. *Ancestry.com.* Web. 2021.<https://www.ancestry.com/family-tree/person/tree/176393814/person/222291669061/hints>.

Department of Commerce and Labor, Bureau of Immigration and Naturalization. "Petition for Naturalization: Williamina Fleming." 1906-1929.   Web. 2021.<https://www.ancestry.com/imageviewer/collections/2361/images/007774867_00338?pId=1826828>.

Donaghe, Harriet Richardson. "Photographic Flashes from Harvard Observatory." *Popular Astronomy* VI.9 (1898): 481-487. Web. 2021.<https://articles.adsabs.harvard.edu//full/1898PA......6..481D/0000481.000.html>.

Editors of Encyclopaedia Britannica. *Hermann Karl Vogel*. 30 Mar 2022. W   e   b   . 2022.<https://www.britannica.com/biography/Hermann-Karl-Vogel >.

"England's   Deathless   Heroes   (Advertisement)."   *The Courier   and   Argus*   (1854):   1.Web.   2020. <https://www.newspapers.com/image/396723297/>.

Fine, Tom. *Harvard College Observatory History in Images*. n.d. Harvard-Smithsonian   Center   for   Astrophysics.   Web. 2021.<https://hea-www.harvard.edu/~fine/Observatory/pages/play.html>.

Fleming, M."A Field for Women's Work in Astronomy." *Astronomy and Astrophysics*   (1893):   683-689.   Web. 2021.<https://wolba.ch/gazette/wp-content/uploads/2017/07/Fleming-Mrs.-M._Astronomy-Astrophysics_Vol.-12_1893_pp.-683-689.pdf>.

Fleming, R. S. *Early Victorian Undergarments*. n.d. Web. 2021.<http://www.katetattersall.com/early-victorian-undergarments-part-4-pantelettes-pantalettes/>.

Fleming, Williamina. "A Photographic Study of Variable Stars." *Annals of the Astronomical Observatory of Harvard College* 1907. Web. 2021.<https://articles.adsabs.harvard.edu/full/1907AnHar..47....1F>.

Fleming, Williamina P. "Journal of Williamina Paton Fleming." 1900. Harvard University - Harvard University Archives. Web. 2021.<https://iiif.lib.harvard.edu/manifests/view/drs:3007384$1i>.

—. "Spectra and Photographic Magnitudes of Stars in Standard Regions." *Annals of the Astronomical Observatory of Harvard College* LXXI.2

(1917):   27-46.
Web.<https://books.google.com/books?id=ha7nAAAAMAAJ&pg=P
A27&lpg=PA27&dq=Spectra+and+Photographic+Magnitudes+of+
Stars+in+Standard+Regions&source=bl&ots=L8WH62JSyR&sig=-
QeyiAA4v0TcRckOJTybxs1ExdY&hl=en&sa=X&ved=0ahUKEwjW
3siTkcHLAhUEVyYKHQ6aBDsQ6AEIIjAB#v=onepage>.

—. "Stars Having Peculiar Spectra." *Annals of the Astronomical Observatory of Harvard College* 56.6 (1912): 165-226. Web. 2021.<https://articles.adsabs.harvard.edu//full/1912AnHar..56..165F /0000184.000.html>.

Fleming, Williamina Paton Stevens. "Spectra and Photographic Magnitudes of Stars in Standard Regions." (1917). Web.<https://nrs.lib.harvard.edu/urn-3:fhcl:2158488?n=19>.

"Form of Records." Vers. UA V 630.229. n.d. *Papers of the Harvard Observatory.* Harvard University Archives. Web. 2021.

Frost, E. B. "Sherburne Wesley Burnham." *Journal of the Royal Astronomical Society of Canada, Vol. 15, p.269* 15 (1921): 269-275. Web. 2022.<https://articles.adsabs.harvard.edu/full/1921JRASC..15..269F >.

GaBany, R. J. *Astrophoto: Fleming's Triangular Wisp by Steve Cannistra.* 24   August   2006.   Web. 2021.<https://www.universetoday.com/519/astrophoto-flemings-tria ngular-wisp-by-steve-cannistra/>.

Geiling, Natasha. "The Women Who Mapped the Universe And Still Couldn't Get Any Respect." *Smithsonian Magazine* (2013). Web. 2021.<https://www.smithsonianmag.com/history/the-women-who- mapped-the-universe-and-still-couldnt-get-any-respect-9287444/>.

Glover, Bill. *CSMinia.* 20 December 2020. Bill Burns. Web. 2021.<https://atlantic-cable.com/Cableships/Minia/>.

Gregersen, Erik. "Mary Somerville". Encyclopedia Britannica, 22 Dec. 2024. Web.20 April 2025. <https://www.britannica.com/biography/Mary-Somerville>.

Guerra, Cristela. "'Women computers' often couldn't use Harvard's telescope. They changed astronomy anyway." *The Boston Globe* 10 August   2017.   Web. 2021.<https://www.bostonglobe.com/lifestyle/2017/08/10/women-c

omputers-held-stars-their-hands/qfLYwpsNZdFNHyiY2igPNJ/story.html>.

Haley, Paul A. "Williamina Fleming and the Harvard College Observatory." *The Antiquarian Astronomer* 11 (2017): 2-32. Web. 2021.

Harvard College Observatory. *Harvard College Observatory observations, logs, instrument readings, and calculations.* n.d. Harvard-Smithsonian Center for Astrophysics. Web. 2020.<https://hollisarchives.lib.harvard.edu/repositories/32/resources/7950>.

"Harvard's Women Astronomers- Work of Mrs. Fleming and Her Associates." *The New York Times* 23 Oct 1904. Web. 2021.<https://www.nytimes.com/1904/10/23/archives/harvards-women-astronomers-work-of-mrs-fleming-and-her-associates.html>.

Harwood, Margaret. "Arthur Seale." *Popular Astronomy* 29.7 (1921): 377-381. Web.2022. <https://adsabs.harvard.edu/full/1921PA.....29..377H>.

Hills, William H. "Excitement of Observers. Eclipse-struck Amateurs Found the Minutes Going Like Seconds--Many Went to Work Breakfastless." *The Boston Globe* 28 May 1900: 1. Web.

—. "Perfect Day. Observers Had Great Success." *The Boston Daily Globe* 28 May1900: 1, 4, 8, 9. Web. 2021.<https://www.newspapers.com/image/428546622/>.

History of Women and the Royal Astronomical Society. n.d. Web. 2021.<https://women.ras.ac.uk/women-and-the-ras/history-of-women-at-the-ras>.

Hoffleit, E. Dorrit. "Pioneering Women in the Spectral Classification of Stars." *Physics in Perspective* 4 (2002): 370-39. Web. 2021.<http://web.science.mq.edu.au/~orsola/MQTeaching/ASTR377/Hoffleit.pdf>.

"Honor American Women." *Monmouth Democrat* 5 Jul 1906: 2. Web. 2021.<https://www.newspapers.com/image/497161192/>.

"Honor American Women." *The Galena Evening Times* 21 Jul 1906: 2. Web.2021. <https://www.newspapers.com/image/142547225/>.

"Honor American Women." *The Red Rock Opinion* 15 Dec 1906: 5. Web. 2021.<https://www.newspapers.com/image/603752408/>.

"Honor American Women." *Daily News-Telegraph* 9 Jul 1906: 4. Web. 2021.<https://www.newspapers.com/image/37520713/>.

"Honor American Women." *Stevens Point Journal* 13 Jul 1906: 3. Web. 2021.<https://www.newspapers.com/image/251222375/>.

"Honor American Women." *St. Joseph Saturday Herald* 28 Jul 1906: 6. Web.2021. <https://www.newspapers.com/image/362306936/>.

"Honor American Women." *The Weekly Advocate* 21 Jul 1906: 3. Web. 2021.<https://www.newspapers.com/image/436648268/>.

Hughes, Stefan. *Catchers of the Light: The Forgotten Lives of the Men and Women Who First Photographed the Heavens*. ArtDeCiel Publishing, 2   0   1   2   . Web.<https://books.google.com/books?id=iZk5OOf7fVYC&dq=hughes+2012+williamina+fleming&lr=&source=gbs_navlinks_s>.

Komar, Marlen. "How 19th-Century Activists Ditched Corsets for One-Piece Long Underwear." *Smithsonian Magazine* 19 Jan 2021. 2021.<https://www.smithsonianmag.com/innovation/how-19th-century-activists-ditched-corsets-for-one-piece-long-underwear-180976774/>.

Lafortune, Keith R. *Women at the Harvard College Observatory, 1877 - 1919: "Women's Work, the "New" Sociality of Astronomy, and Scientific L   a   b   o   r   .* 2001.Web.<https://www.proquest.com/openview/d75c2f60d1b11e414f64e4e3226f757b/1?pq-origsite=gscholar&cbl=18750&diss=y>.

Mack, P. E. "Strategies and Compromises - Women in Astronomy at Harvard College Observatory 1870-1920." *Journal for the History of Astronomy, Vol.21,NO. 1/FEB, P. 65, 1990* 21.1 (1990): 65-76. Web.<https://articles.adsabs.harvard.edu/full/1990JHA....21...65M>.

Massachusetts Bay Transportation Authority. *The History of the T.* n.d. MBTA. Web. 2022.<mbta.com/history>.

Massachusetts Department of Labor and Industries, Bureau of Statistics. "Annual report on the statistics of labor. 1884." Vers. 2019-11-28 13:33 UTC.   July   1884.   *BabelHathiTrust.org.*   Web. 2021.<https://babel.hathitrust.org/cgi/pt?id=coo.31924054162171&view=1up&seq=466&skin=2021>.

"Massachusetts, U. S. , State and Federal Naturalization Records, 1798-1950, for Williamina Fleming." 1902. *Ancestry.com.*

Commonwealth of Massachusetts. Web.2021. <https://www.ancestry.com/discoveryui-content/view/1826828:2361 >.

"Massachusetts, U.S., Death Records, 1841-1915." n.d. *Ancestry.com*. Ancestry.com Operations. Web. 2021.<https://www.ancestry.com/imageviewer/collections/2101/images/41262_b131967-00485?pId=8519454>.

"Massachusetts, U.S., State and Federal Naturalization Records, 1798-1950." n.d. *Ancestry.com*. Web. 2021.<https://www.ancestry.com/family-tree/person/tree/176393814/person/222296869168/hints>.

McEachern, Maria C. "Every Star Speaks for Itself: Williamina Fleming and the Work of the Harvard College Observatory." *Galactic Gazette: A Blog from the staff of Wolbach Library* (2013). Web. 2021.<https://wolba.ch/gazette/every-star-speaks-for-itself/>.

McGrath, Alex. *The First Computer: Williamina Fleming and the Horsehead Nebula*. 3 July 2017. Harvard and Smithsonian Center for Astrophysics. Web. 2021.<https://wolba.ch/gazette/the-first-computer-williamina-fleming-and-the-horsehead-nebula/>.

Memorial Page for Williamina Paton Stevens Fleming. 9 Sep 2010. Find a Grave. Web. 2020.<https://www.findagrave.com/memorial/58437055/williamina-paton-fleming>.

"Miscellaneous Quarto Publications, 1877-1896." 4 (1896). Web. 2021.<https://books.google.com/books?id=Z-sRAAAAYAAJ&pg=PA165&lpg=PA165&dq=History+and+Progress+of+the+Henry+Draper+Memorial+During+the+Years+1886-91&source=bl&ots=K8Lzf AFCH0&sig=ACfU3U1iYGTBUMrb0C1RJuYiEG6iu1S7jQ&hl=en&sa=X&ved=2ahUKEwjUlNnY6obxAhXNGs0KHZ_ABms>.

"Mrs. Elizabeth Wadsworth Pickering (Obituary) 1906." *Cambridge Tribune* XXIX.27 (1906): 4. Web. 2021.<https://cambridge.dlconsulting.com/?a=d&d=Tribune19060901-01.2.34&e=-------en-20--1--txt-txIN------->.

"Mrs. Williamina P. Fleming Dead: Curator at Harvard Observatory." *Boston Daily Globe* 22 May 1911: 16. Web. 2021.<https://www.newspapers.com/image/430662063/>.

Musacchio, John S. Cameron and Jacqueline Marie. "Sarah Frances Whiting and the "photography of the invisible"." *Physics Today* 73.8 (2020): 26. Web. 2022.<https://physicstoday.scitation.org/doi/full/10.1063/PT.3.4545 >.

Myers, A. Wallis. "On Board a Cable Hospital (1902)." *The Windsor Magazine* (1902). Web. 2021.<https://atlantic-cable.com/Article/1902CableHospital/index.h tm>.

National Archives and Records Administration. "U.S., Passport Applications, 1795-1925[database on-line]." 1920. *Ancestry.com.* Web. 2021.<https://www.ancestry.com/discoveryui-content/view/606764:1 174?tid=176393814&pid=222295571783&queryId=e69fb91274c0e 8fe46c32ab454b1a426&_phsrc=wmL114&_phstart=successSource>.

—. "United States of America Twelfth Census of the United States, 1900." 1900. *Ancestry.com.* Web. 2021.<https://www.ancestry.com/imageviewer/collections/7602/ima ges/4113830_00794?pId=79188963>.

National Archives at Boston; Waltham, Massachusetts. "Records of the Immigration and Naturalization Service, 1787-2004." n.d. *Ancestry.com.* Web. 2021.<https://www.ancestry.com/imageviewer/collections/2361/ima ges/007137226_00358?treeid=176393814&personid=222295571783 &hintid=&queryId=e69fb91274c0e8fe46c32ab454b1a426&usePUB =true&_phsrc=wmL115&_phstart=successSource&usePUBJs=true& _ga=2.181748775.1086561509.>.

National Archives at Washington, D.C. "Massachusetts, U.S., Arriving Passenger and Crew Lists, 1820-1963." n.d. *Ancestry.com.* Web. 2021.<https://www.ancestry.com/imageviewer/collections/8745/ima ges/MAM277_105-0077?treeid=176393814&personid=2222955717 83&rc=&usePUB=true&_phsrc=wmL119&_phstart=successSource &pId=3554877>.

National Library of Scotland. *Williamina Fleming: Astronomer, 1857-1911.* n.d. Web. 2021.<https://www.nls.uk/learning-zone/science-and-technology/women-scientists/williamina-fleming>.

"New Hampshire, U.S., Death and Disinterment Records, 1754-1947 for Mary Stevens." n.d. *Ancestry.com.* Web. 2021.<https://www.ancestry.com/imageviewer/collections/5242/images/41267_311863-01668?pId=379912>.

"New Names of Minor Planets." *WGSBN Bulletin* 17 Jan 2022: 5. Web. 25 Mar 2022.<https://www.wgsbn-iau.org/files/Bulletins/V002/WGSBNBull_V002_001.pdf>.

"New Star Discovered by Woman Astronomer." *Fort Worth Star-Telegram* 15 Oct 1910: 3. Web. 2021. <https://www.newspapers.com/image/634054210/>.

"New Star Discovered: Mrs. Williamina Fleming Adds to Her Astronomical Record." *Pittsburgh Post-Gazette* 15 Oct 1910: 1. Web. 2021.<https://www.newspapers.com/image/85713579/>.

Newman, Alex. *A team of women is unearthing the forgotten legacy of Harvard's women 'computers'.*27 July 2017. PRX, GBH. Web. 2021.<https://theworld.org/stories/2017-07-27/team-women-are-unearthing-forgotten-legacy-harvard-s-women-computers>.

"Noted Woman Guest of Son in the City." *The Salt Lake Tribune* 22 Aug. 1909: 20.Web. 2021. <https://www.newspapers.com/image/76420676/>.

Pennie, Len. "Scots Word of the Day." *Lentil Pentil@Lenniesaurus.* n.d. Twitter. 2021.

PHaEDRA Summer Art Series, #1: Measure. Perf. Anna Von Mertens. 8 June 2021. Live Art Presentation via Web. 8 June 2021.

PHaEDRA Summer Art Series, #2: Joyce Van Dyke. Perf. Joyce Van Dyke. 16 July 2021. Live Performance via Web. 16 July 2021.

Pickering, E. C. and W. P. Fleming. "Photographic observations of variable stars during the years 1886 to 1905 forming a part of the Henry Draper memorial." *Annals of Harvard College Observatory, vol. 47, pp.115-280* 47 (1912): 115-280.Web.

2021.<https://articles.adsabs.harvard.edu/full/1912AnHar..47..115F>.

Pickering, E. C. "Detection of New Nebulae by Photography." *Annals of Harvard College Observatory* XVIII.VI (1890): 113-117. Web. 2021.<https://articles.adsabs.harvard.edu/full/1890AnHar..18..113P>.

—. "New Star in Sagittarius." *Harvard Astronomical Bulletin* 4 Oct 1910. Web.2021.<https://books.google.com/books?id=dQlLAAAAYAAJ&pg=PA563&lpg=PA563&dq=new+star+in+sagittarius+astronomical+bulletin+426&source=bl&ots=NGPL9CKOUN&sig=ACfU3U01sgxC28sA_Cn-0MGsoc2_1VEK3g&hl=en&sa=X&ved=2ahUKEwjZtLzOkeD2AhVImeAKHXT6A3kQ6AF6BAgCEAM#v=onepage&q=>.

—. "Nova Arae." *Harvard Astronomical Bulletin* 1910. Web. 2021.<https://books.google.com/books?id=dQlLAAAAYAAJ&pg=PA563&lpg=PA563&dq=new+star+in+sagittarius+astronomical+bulletin+426&source=bl&ots=NGPL9CKOUN&sig=ACfU3U01sgxC28sA_Cn-0MGsoc2_1VEK3g&hl=en&sa=X&ved=2ahUKEwjZtLzOkeD2AhVImeAKHXT6A3kQ6AF6BAgCEAM#v=onepage&q=>.

Pickering, Edward C. "Sixty-sixth Annual Report of the Director of the Astronomical Observatory of Harvard College for the year ending September 30, 1911." *Harvard College Observatory Annual Report* 66 (1911): 3-9. Web. 2021.<https://articles.adsabs.harvard.edu/full/1912HarAR..66....3P>.

—."Williamina Paton Fleming." *The Harvard Graduates Magazine* September 1911: 49-51. Web. 2021.<https://www.google.com/books/edition/The_Harvard_Graduates_Magazine/4dFNAAAAMAAJ>.

Pickering, Edward Charles. *An Investigation in Stellar Photography Conducted at the Harvard College Observatory*. Cambridge, Mass.: University Press, 1886. Web. 2021.<https://books.google.com/books?hl=en&lr=&id=oIDvAAAAMAAJ&oi=fnd&pg=PA177&dq=annals+of+harvard+college+observatory&ots=W7-5X40M-x&sig=FZLP7AZpUbDQlvx5wdNWRGsBtXY#v=onepage&q=fleming&f=false>.

—."Preparation and Discussion of the Draper Catalogue, Vol. 26." *Annals of the Astronomical Observatory of Harvard College* XXVI, Part 1 (1891). Web. 2021.<https://books.google.com/books?hl=en&lr=&id=B6jnAAAA MAAJ&oi=fnd&pg=PR9&dq=annals+of+harvard+college+observat ory&ots=JWmOWlG8hW&sig=mPuBfZhB5g0ty4I_WTdZW3D18n s#v=snippet&q=1&f=false>.

—. *Report on the work of the observatory for 1887. Observations included positions and magnitudes for the satellites of Mars and Saturn.* Cambridge, Mass.: Harvard College Observatory Papers, vol. 1, pp.3.1-3.36, 1877. Web. 2021.<https://articles.adsabs.harvard.edu/full/1877HarPa...1....3P>.

—. "Variable Stars of Long Period." *Pamphlets on Astronomy* 7.31 (1891). Web.2021.<https://books.google.com/books?hl=en&lr=&id=XnPm 5RgwDIoC&oi=fnd&pg=PA1&dq=annals+of+harvard+college+obs ervatory&ots=4DPcj451q8&sig=Bl1HO1RjHLJ7Mre37koEviPVKv Y#v=onepage&q=fleming&f=false>.

Pittendreigh, W. Maynard. "Williamina Fleming: A Pioneering Woman in Astronomy." *Reflector*73.2 (2021): 22-23. Web. 2021.<https://www.astroleague.org/files/reflector/Reflector%20Marc h%202021%20Final%20spreads_0.pdf>.

Reed, Helen Leah. "Women's Work at the Harvard Observatory." *New England Magazine* (1925): 165-176. Web. 2021.<https://id.lib.harvard.edu/curiosity/expeditions-and-discoveries /38-990086499030203941>.

Ritchie, Gayle. "Williamina Fleming: The Dundee-born Victorian maid who mapped the heavens." *Evening Telegraph* (2020). Web. 2021.<http://www.eveningtelegraph.co.uk/fp/williamina-fleming-the -dundee-born-victorian-maid-who-mapped-the-heavens/>.

"Robert Stevens, Carver, Gilder, Printseller, & Artists, Colourman (Advertisement)." *The Courier and Argus* (1856): 1. Web. 2021.<https://www.newspapers.com/image/411017553/>.

Rossiter, Margaret W. ""Women's Work" in Science, 1880-1910." *Isis* Sep 1980: 381-398. Web. 2021. <https://www.jstor.org/stable/230118>.

Saunders, Charles H. "Cambridge Commons." 1896. *The Cambridge of Eighteen Hundred Ninety-six: a picture of the city and its industries fifty years after its corporation.* Ed. Arthur Gilman. Nineteenth Century American. Web. January2022. <perseus.tufts.edu/hopper>.

"Scotland Select Births and Baptisms, 1564-1950 [database on-line]." 2014. *Ancestry.com.* Ancestry.com Operations. Web. 2021.

Scotland, Select Births and Baptisms, 1564-1950. n.d. Web. 2020.<https://www.ancestry.com/family-tree/person/tree/176393814 /person/222291669062/facts?msg=ntm&msgParams=%7c1%7c1%7c &mpid=222291669062&nec=0&mdbid=7602&mrpid=6162050>.

"She Discovers Stars." *The Elyria Reporter* 23 Sep 1905: 3. Web. 2021.<https://www.newspapers.com/image/11600597/>.

"She Discovers Stars." *Altoona Tribune* 2 Nov 1905: 9. Web. 2021.<https://www.newspapers.com/image/61513107/>.

"She Discovers Stars." *Archbald Citizen* 23 Sep 1905: 3. Web. 2021.<https://www.newspapers.com/image/640025585/>.

"She Discovers Stars." *The DeSoto County News* 29 Sep 1905: 12. Web.2021. <https://www.newspapers.com/image/75702707/>.

"She Discovers Stars." *The Sentinel* 2 Oct 1905: 7. Web. 2021.<https://www.newspapers.com/image/344605867/>.

"She Discovers Stars." *The McCook Tribune* 29 Sep 1905: 6. Web. 2021.<https://www.newspapers.com/image/98873414/>.

"She Discovers Stars." *The Beatrice Express* 27 Oct 1905: 4. Web. 2021.<https://www.newspapers.com/image/506585048/>.

"She Discovers Stars." *The Bucyrus Evening Telegraph* 16 Sep 1905: 6.Web. 2021. <https://www.newspapers.com/image/600842135/>.

Shears, Jeremy. "Thomas Hinsley Astbury: from an English market town schoolroom to the internal constitution of the stars." *Journal of the British Astronomical Association* 124.2 (n.d.): 81-91. Web. 2021.<https://articles.adsabs.harvard.edu/full/2014JBAA..124...81S> .

Simmons, Mike. *A New Era in Solar Research.* 2021. Mount Wilson Institute.   Web. 2022.<https://www.mtwilson.edu/entering-a-new-era-in-solar-researc h/>.

Sobel, Dava. *The Glass Universe: How the Ladies of the Harvard Observatory Took the Measure of the Stars.* New York: Penguin Random House, 2016. Book.

Sponberg, Brant L. and David H. DeVorkin. *The Origin of the AAS*. 2019. Astronomical Society of the Pacific. Web. 2022. <https://aas.org/about/origins_aas>.

"Taking Observations: The Transit of Venus Watched by Hundreds." *The Boston Globe* 6 Dec 1882: 8. Web. 2021.<https://www.newspapers.com/image/428499611/>.

The Editors of Encyclopaedia Britannica. "Caroline Herschel". Encyclopedia Britannica, 12 Mar. 2025. Web.30 April 2025. <https://www.britannica.com/biography/Caroline-Lucretia-Herschel>.

"The Woman Who Found Fame in the Heavens." *The Oregon Daily Journal* 27 Nov 1910: 65. Web. 2021. <https://www.newspapers.com/image/79473858/>.

Turner, Herbert Hall. "Obituary Notice : Honorary Member :- Fleming, Williamina Paton." *Monthly Notices of the Royal Astronomical Society* 72(1911): 261-264. Web. 2021.<https://adsabs.harvard.edu/full/1912MNRAS..72..261.>.

U.S. Circuit Court, District of Massachusetts. "U.S., Naturalization Records Indexes, 1794-1995." 1906-1911. *Ancestry.com.* Web. 2021.<https://www.ancestry.com/imageviewer/collections/1192/images/M1545_10-1241?pId=7669>.

"U.S. Naturalization Records Indexes, 1794-1995." n.d. *Ancestry.com.* Web.2021.<https://www.ancestry.com/imageviewer/collections/1192/images/M1545_10-1241?pId=7669>.

"U.S. Passport Applications for 1795-1925 for Annie Jump Cannon." n.d. *Ancestry.com.* Web. 2022.<https://www.ancestry.com/imageviewer/collections/1174/images/USM1490_1821-0759?treeid=&personid=&rc=&usePUB=true&_phsrc=wmL179&_phstart=successSource&pId=65995>.

Van Dyke, Joyce. *The Women Who Mapped the Stars*. Dir. Jessica Ernst. Central Square Theatre, Cambridge. Recorded performance and playwright talk. July 2021. Performance.

Wehrey, Catherine. "Women Computing the Stars." *Verso: The Blog of the Huntington Library, Art Museum, and Botanical Gardens* (2015). Web. 2021.<https://www.huntington.org/verso/2015/09/women-computing-stars>.

Wilson, H. C. "The Fourth Conference of the International Union for Co-operation in Solar Research." *Popular Astronomy* 18 (1910): 489-503. Web. 2021.<https://articles.adsabs.harvard.edu/full/1910PA.....18..489W>.

Wolbach Library. *Administrative Work*. 2020. Wolbach Library. Web. 2021.<https://library.cfa.harvard.edu/administrative-work>.

—. *The First Computer: Williamina Fleming and the Horsehead Nebula*. 3 Jul 2017. Harvard and Smithsonian Center for Astrophysics. Web. 2021.

—. *Work With Variable Stars and Nebulae*. 2020. Wolbach Library. Web. 2021.<https://library.cfa.harvard.edu/fleming-work>.

Wolf-Rayet Star. n.d. Swinburne University of Technology. Web. 2022.<https://astronomy.swin.edu.au/cosmos/w/wolf-rayet+star>.

"Woman Astronomer." *The Virginia Enterprise* 21 Sep 1906: 2. Web. 2020.<newspapers.com/image/83884586/>.

"Woman Astronomer." *Coffeyville Daily Record* 23 Oct 1906: 3. Web. 2021.<https://www.newspapers.com/image/418140498/>.

"Woman Astronomer." *The Schuyler Sun* 28 Sep 1906: 5. Web. 2021.<https://www.newspapers.com/image/555907094/>.

"Woman Astronomer." *The Daily Item* 21 Sep 1906: 3. Web. 2021.<https://www.newspapers.com/image/366890045/>.

"Woman Astronomer." *The Stark News* 22 Sep 1906: 3. Web. 2021.<https://www.newspapers.com/image/486633896/>.

"Woman Astronomer." *Garrison Argus* 20 Sep 1906: 7. Web. 2021.<https://www.newspapers.com/image/690850540/>.

"Woman Astronomer." *The Frederick Leader* 20 Sep 1906: 7. Web. 2021.<https://www.newspapers.com/image/584990315/>.

"Woman Astronomer." *The Lincoln Herald* 21 Sep 1906: 5. Web. 2021.<https://www.newspapers.com/image/309601556/>.

"Woman Astronomer." *The Journal* 20 Sep 1906: 2. Web. 2021.<https://www.newspapers.com/image/322914137/>.

"Woman Astronomer." *Brookville Headlight* 21 Sep 1906: 3. Web. 2021.<https://www.newspapers.com/image/487171938/>.

"Woman Astronomer." *Williamsburg Star* 21 Sep 1906: 6. Web. 2021.<https://www.newspapers.com/image/427774540/>.

"Woman Astronomer Discovered Six Out of Nine New Stars." *The Boston Globe* 26 Aug 1906: 51. Web. 2021.<https://www.newspapers.com/image/430832660/>.

"Women in Astronomy: Fair Sex Has Made a Remarkable Record in Study of Stars." *Baltimore Sun* 7 Nov 1910: 12. Web. 2021.<https://www.newspapers.com/image/371647572/>.

"World's Greatest Woman Astronomer." *The Philadelphia Inquirer* 4 Sep 1905:6. Web. 2021. <https://www.newspapers.com/image/168420795/>.

"Yale Defeats Harvard in a Game of Foot Ball." *The Boston Globe* 26 Nov 1882: 8.Web. 2021. <https://www.newspapers.com/image/428498694/>.